Pearson Australia
(a division of Pearson Australia Group Pty Ltd)
459–471 Church Street, Level 1, Building B, Richmond, Victoria 3121
PO Box 23360, Melbourne, Victoria 8012

www.pearson.com.au

First published 2018 by Pearson Australia
2027 2026 2025 2024
1 2 3 4 5 27 26 25 24

Lead Publishers: Misal Belvedere and Malcolm Parsons
Project Manager: Michelle Thomas
Production Editors: Laura Pietrobon and Virginia O'Brien
Lead Development Editor: Fiona Cooke
Content Developer: Bryonie Scott
Development Editor: Naomi Campanale
Editor: Sam Trafford
Designers: Anne Donald, iEnergizer Aptara Ltd
Rights & Permissions Editor: Samantha Russell-Tulip
Senior Publishing Services Analyst: Rob Curulli
Proofreader: Diane Fowler
Illustrator: DiacriTech
Printed in Australia by Pegasus Media and Logistics

ISBN 978 1 4886 1936 6

Pearson Australia Group Pty Ltd ABN 40 004 245 943

Attributions

Cover image: Daniel Chen/Unsplash

The following abbreviations are used in this list: t = top, l = left, r = right, c = centre.

Alamy Stock Photo: Stephen Barnes, p. 4; Carsten Leuzinger/imageBROKER, p. 100–2; Photo Researchers/Science History Images, p. 155–7.

ANSTO: p. 209.

Getty Images: Quinn Rooney, p. 6.

Malcolm Cross: p. 105b.

NASA Images: NASA/Johnson Space Centre, p. 51–3.

Science Photo Library: p. 146; Andrew Lambert Photography, pp. 105t, 158cl, 158cr; Alex Cherney, Terrastro.com, p. 1–3; FERMI National Accelerator Laboratory, p. 188; Jo Lomberg, p. 171; Royal Astronomical Society, p. 44.

Shutterstock: Everett Historical, p. 16; muzsy, p. 5; noolwlee, p. 94; Surakit, pp. 8–9.

Unsplash: Daniel Chen, p. i.

Some of the images used in *Pearson Physics 12 New South Wales Skills and Assessment* book might have associations with deceased Indigenous Australians. Please be aware that these images might cause sadness or distress in Aboriginal or Torres Strait Islander communities.

Practical activities
All practical activities, including the illustrations, are provided as a guide only and the accuracy of such information cannot be guaranteed. Teachers must assess the appropriateness of an activity and take into account the experience of their students and the facilities available. Additionally, all practical activities should be trialled before they are attempted with students and a risk assessment must be completed. All care should be taken whenever any practical activity is conducted: appropriate protective clothing should be worn, the correct equipment used, and the appropriate preparation and clean-up procedures followed. Although all practical activities have been written with safety in mind, Pearson Australia and the authors do not accept any responsibility for the information contained in or relating to the practical activities, and are not liable for any loss and/or injury arising from or sustained as a result of conducting any of the practical activities described in this book.

Contents

Contents

Module 7: The nature of light

Module 8: From the universe to the atom

How to use this book

The *Pearson Physics 12 New South Wales Skills and Assessment* book takes an intuitive, self-paced approach to science education that ensures every student has opportunities to practise, apply and extend their learning through a range of supportive and challenging activities. While offering opportunities for reinforcement of key concepts, knowledge and skills, these activities enable flexibility in the approach to teaching and learning.

Explicit scaffolding makes learning objectives clear, and there are regular opportunities for student reflection and self-evaluation at the end of individual activities throughout the book. Students are also guided in self-reflection at the end of each module. There are rich opportunities to take the content further with the explicit coverage of Working scientifically skills and key knowledge in the depth studies.

This resource has been written to the new Stage 6 Syllabus for New South Wales Physics and addresses the final four modules of the syllabus. Each module consists of five main sections:

- key knowledge
- worksheets
- practical activities
- depth study
- module review questions.

Explore how to use this book below.

Physics toolkit

The Physics toolkit supports development of the skills and techniques needed to undertake practical investigations, secondary-sourced investigations and depth studies, and covers examination techniques and study skills. It also includes checklists, models, exemplars and scaffolded steps. The toolkit can serve as a reference tool, to be consulted as needed.

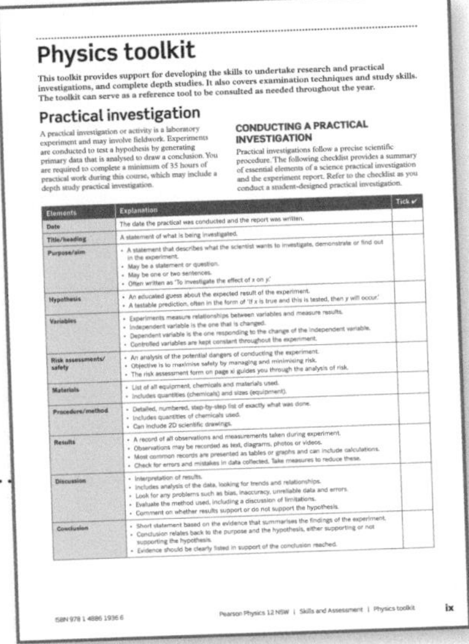

Physics toolkit

This toolkit provides support for developing the skills to undertake research and practical investigations, and complete depth studies. It also covers examination techniques and study skills. The toolkit can serve as a reference tool to be consulted as needed throughout the year.

Practical investigation

A practical investigation or activity is a laboratory experiment and may involve fieldwork. Experiments are conducted to test a hypothesis by generating primary data that is analysed to draw a conclusion. You are required to complete a minimum of 35 hours of practical work during this course, which may include a depth study practical investigation.

CONDUCTING A PRACTICAL INVESTIGATION

Practical investigations follow a precise scientific procedure. The following checklist provides a summary of essential elements of a science practical investigation and the experiment report. Refer to the checklist as you conduct a student-designed practical investigation.

Elements	Explanation	Tick ✓
Date	The date the practical was conducted and the report was written.	
Title/heading	A statement of what is being investigated.	
Purpose/aim	• A statement that describes what the scientist wants to investigate, demonstrate or find out in the experiment. • May be a statement or question. • May be one or two sentences. • Often written as 'To investigate the effect of x on y.'	
Hypothesis	• An educated guess about the expected result of the experiment. • A testable prediction, often in the form of 'If x is true and this is tested, then y will occur.'	
Variables	• Experiments measure relationships between variables and measure results. • Independent variable is the one that is changed. • Dependent variable is the one responding to the change of the independent variable. • Controlled variables are kept constant throughout the experiment.	
Risk assessments/safety	• An analysis of the potential dangers of conducting the experiment. • Objective is to maximise safety by managing and minimising risk. • The risk assessment form on page xi guides you through the analysis of risk.	
Materials	• List of all equipment, chemicals and materials used. • Includes quantities (chemicals) and sizes (equipment).	
Procedure/method	• Detailed, numbered, step-by-step list of exactly what was done. • Includes quantities of chemicals used. • Can include 2D scientific drawings.	
Results	• A record of all observations and measurements taken during experiment. • Observations may be recorded as text, diagrams, photos or videos. • Most common records are presented as tables or graphs and can include calculations. • Check for errors and mistakes in data collected. Take measures to reduce these.	
Discussion	• Interpretation of results. • Includes analysis of the data, looking for trends and relationships. • Look for any problems such as bias, inaccuracy, unreliable data and errors. • Evaluate the method used, including a discussion of limitations. • Comment on whether results support or do not support the hypothesis.	
Conclusion	• Short statement based on the evidence that summarises the findings of the experiment. • Conclusion relates back to the purpose and the hypothesis, either supporting or not supporting the hypothesis. • Evidence should be clearly listed in support of the conclusion reached.	

ISBN 978 1 4886 1936 6 Pearson Physics 12 NSW | Skills and Assessment | Physics toolkit ix

Module opener

Each book is divided to follow the four modules of the syllabus, with the module opener linking the module content to the syllabus.

Key knowledge

Each module begins with a key knowledge section. The key knowledge consists of a set of succinct summary notes that cover the key knowledge set out in each module of the syllabus. This section is highly illustrative and written in a straightforward style to assist students of all reading abilities. Key terms are bolded for ease of navigation. It also serves as a ready reference for completing the worksheets and practical activities.

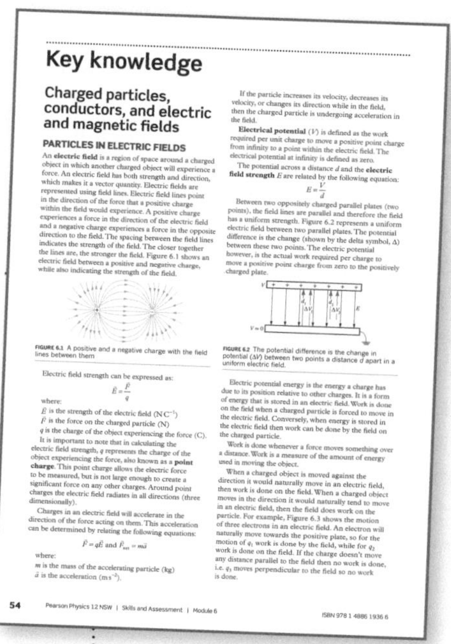

Key knowledge

Charged particles, conductors, and electric and magnetic fields

PARTICLES IN ELECTRIC FIELDS

An **electric field** is a region of space around a charged object in which another charged object will experience a force. An electric field has both strength and direction, which makes it a vector quantity. Electric fields are represented using field lines. Electric field lines point in the direction of the force that a positive charge within the field would experience. A positive charge experiences a force in the direction of the electric field and a negative charge experiences a force in the opposite direction to the field. The spacing between the field lines indicates the strength of the field. The closer together the lines are, the stronger the field. Figure 6.1 shows an electric field between a positive and negative charge, while also indicating the strength of the field.

FIGURE 6.1 A positive and a negative charge with the field lines between them

Electric field strength can be expressed as:

$$\vec{E} = \frac{\vec{F}}{q}$$

where:

$\vec{E}$ is the strength of the electric field ($N C^{-1}$)

$\vec{F}$ is the force on the charged particle (N)

q is the charge of the object experiencing the force (C).

It is important to note that in calculating the electric field strength, q represents the charge of the object experiencing the force, also known as a **point charge**. This point charge allows the electric force to be measured, but is not large enough to create a significant force on any other charges. Around point charges the electric field radiates in all directions (three dimensionally).

Charges in an electric field will accelerate in the direction of the force acting on them. This acceleration can be determined by relating the following equations:

$$\vec{F} = q\vec{E} \text{ and } \vec{F}_{net} = m\vec{a}$$

where:

m is the mass of the accelerating particle (kg)

$\vec{a}$ is the acceleration ($m s^{-2}$).

If the particle increases its velocity, decreases its velocity, or changes its direction while in the field, then the charged particle is undergoing acceleration in the field.

Electrical potential (V) is defined as the work required per unit charge to move a positive point charge from infinity to a point within the electric field. The electrical potential at infinity is defined as zero.

The potential across a distance d and the **electric field strength** E are related by the following equation:

$$E = \frac{V}{d}$$

Between two oppositely charged parallel plates (two points), the field lines are parallel and therefore the field has a uniform strength. Figure 6.2 represents a uniform electric field between two parallel plates. The potential difference is the change (shown by the delta symbol, Δ) between these two points. The electric potential however, is the actual work required per charge to move a positive point charge from zero to the positively charged plate.

FIGURE 6.2 The potential difference is the change in potential (ΔV) between two points a distance d apart in a uniform electric field.

Electric potential energy is the energy a charge has due to its position relative to other charges. It is a form of energy that is stored in an electric field. Work is done on the field when a charged particle is forced to move in the electric field. Conversely, when energy is stored in the electric field then work can be done by the field on the charged particle.

Work is done whenever a force moves something over a distance. Work is a measure of the amount of energy used in moving the object.

When a charged object is moved against the direction it would naturally move in an electric field, then work is done on the field. When a charged object moves in the direction it would naturally tend to move in an electric field, then the field does work on the particle. For example, Figure 6.3 shows the motion of three electrons in an electric field. An electron will naturally move towards the positive plate, so for the motion of q_1 work is done by the field, while for q_2 work is done on the field. If the charge doesn't move any distance parallel to the field then no work is done, i.e. q_3 moves perpendicular to the field so no work is done.

54 Pearson Physics 12 NSW | Skills and Assessment | Module 6 ISBN 978 1 4886 1936 6

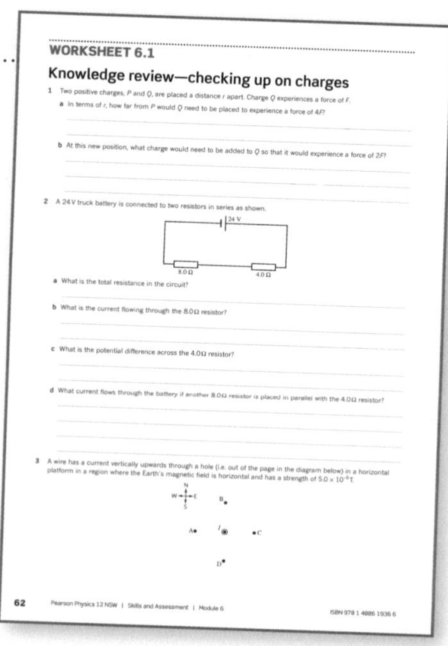
WORKSHEET 6.1
Knowledge review—checking up on charges

Worksheets

A diverse offering of instructive and self-contained worksheets is included in each module. Common to all modules are the initial 'Knowledge review' worksheet to activate prior knowledge, a 'Literacy review' worksheet to explicitly build understanding and application of scientific terminology, and finally a 'Thinking about my learning' worksheet, which students can use for reflection and self-assessment. Other worksheet types provide opportunities to revise, consolidate and further student understanding.

All worksheets function as formative assessment and are clearly aligned to the syllabus. A range of questions building from foundation to challenging are included within worksheets.

Practical activities

Practical activities give students the opportunity to complete practical work related to the various themes covered in the syllabus. All practical activities referenced in outcomes within the syllabus have been covered. Across the suite of practical activities, students have opportunities to design, conduct, evaluate, gather and analyse data, appropriately record results and prepare evidence-based conclusions. Students have opportunities to evaluate safety and risk, and identify any potential hazards.

Each practical activity includes a suggested duration. Along with the depth studies, the practical activities meet the 35 hours of practical work mandated at Year 12 in the syllabus. Where there is key knowledge that will support the completion of a practical activity, students are referred back to it.

Like the worksheets, the practical activities include a range of questions, building from foundation to challenging.

Depth study

Each module contains one suggested depth study. The depth studies allow further development of one or more concepts found within or inspired by the syllabus. They allow students to acquire a depth of understanding and take responsibility for their own learning, and promote differentiation and engagement.

Each depth study allows for the demonstration of a range of Working scientifically skills, with all depth studies addressing the Working scientifically outcomes of Questioning and predicting, and Communicating. A minimum of two additional Working scientifically skills and at least one Knowledge and understanding outcome are also assessed.

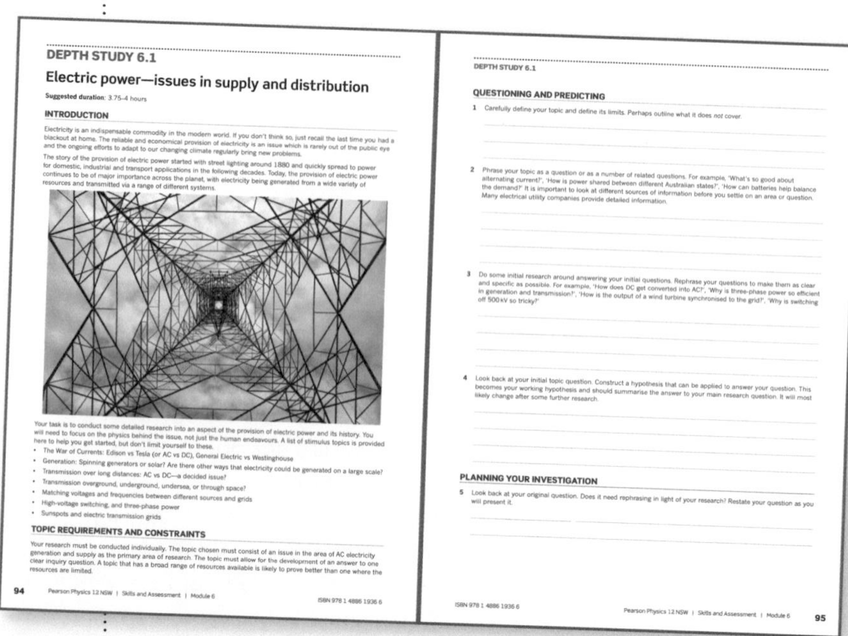
DEPTH STUDY 6.1
Electric power—issues in supply and distribution

 ISBN 978 1 4886 1936 6

Module review questions

Each module finishes with a comprehensive set of questions, consisting of multiple choice and short answer, which helps students to draw together their knowledge and understanding and apply it to these styles of question.

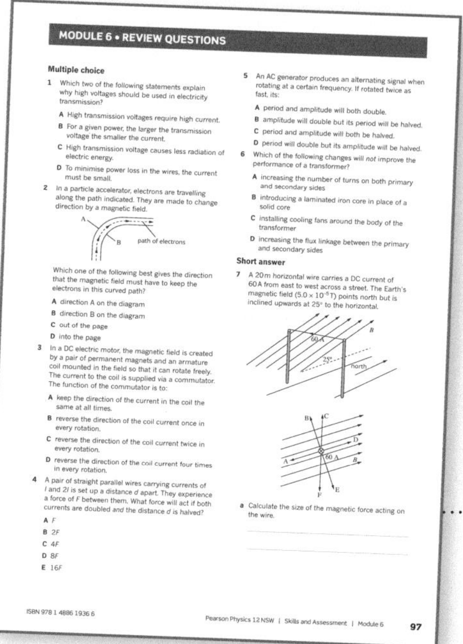

MODULE 6 • REVIEW QUESTIONS

Multiple choice

1 Which two of the following statements explain why high voltages should be used in electricity transmission?

A High transmission voltages require high current.
B For a given power, the larger the transmission voltage the smaller the current.
C High transmission voltage causes less radiation of electric energy.
D To minimise power loss in the wires, the current must be small.

2 In a particle accelerator, electrons are travelling along the path indicated. They are made to change direction by a magnetic field.

Which one of the following best gives the direction that the magnetic field must have to keep the electrons in this curved path?

A direction A on the diagram
B direction B on the diagram
C out of the page
D into the page

3 In a DC electric motor, the magnetic field is created by a pair of permanent magnets and an armature coil mounted in the field so that it can rotate freely. The current to the coil is supplied via a commutator. The function of the commutator is to:

A keep the direction of the current in the coil the same at all times.
B reverse the direction of the coil current once in every rotation.
C reverse the direction of the coil current twice in every rotation.
D reverse the direction of the coil current four times in every rotation.

4 A pair of straight parallel wires carrying currents of I and $2I$ is set up a distance d apart. They experience a force of F between them. What force will act if both currents are doubled *and* the distance d is halved?

A F
B $2F$
C $4F$
D $8F$
E $16F$

5 An AC generator produces an alternating signal when rotating at a certain frequency. If rotated twice as fast, its:

A period and amplitude will both double.
B amplitude will double but its period will be halved.
C period and amplitude will both be halved.
D period will double but its amplitude will be halved.

6 Which of the following changes will *not* improve the performance of a transformer?

A increasing the number of turns on both primary and secondary sides
B introducing a laminated iron core in place of a solid core
C installing cooling fans around the body of the transformer
D increasing the flux linkage between the primary and secondary sides

Short answer

7 A 20 m horizontal wire carries a DC current of 60 A from east to west across a street. The Earth's magnetic field (5.0×10^{-5} T) points north but is inclined upwards at 25° to the horizontal.

a Calculate the size of the magnetic force acting on the wire.

ISBN 978 1 4886 1936 6

Pearson Physics 12 NSW | Skills and Assessment | Module 6 97

Rating my learning

This feature is an innovative tool that appears at the bottom of the final page of most worksheets and all practical activities. It provides students with the opportunity for self-reflection and self-assessment. It encourages them to look ahead to how they can continue to improve, and it helps them to identify focus areas for further skill and knowledge development.

The teacher may choose to use student responses to the 'Rating my learning' feature as a formative assessment tool. At a glance, teachers can assess which topics and which students need intervention for improvement.

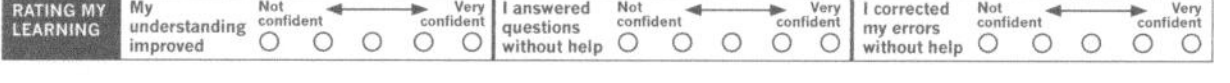

Icons and features

The 2018 New South Wales Physics Stage 6 Syllabus Learning Across the Curriculum content is addressed and identified.

The **safety icon** highlights significant hazards, indicating caution is needed.

The **safety glasses icon** highlights that protective eyewear is to be worn during the practical activity.

Highlight boxes focus students' attention on important information such as key definitions, formulae and summary points.

Consult the manuals for your electronic equipment or see your teacher for the options to set these values with your equipment and software.

Teacher Support

Comprehensive answers and fully worked solutions for all worksheets, practical activities, depth studies and module review questions are provided via the *Pearson Physics 12 New South Wales Teacher Support*. An editable suggested assessment rubric for depth studies is also provided.

Pearson Physics 12 New South Wales

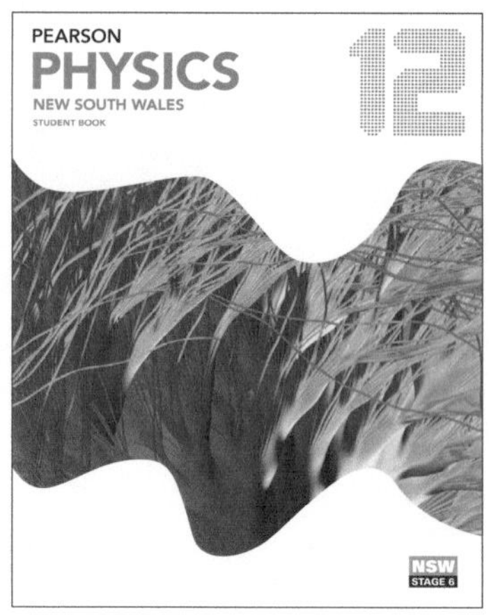

Student Book

Pearson Physics 12 New South Wales has been written to fully align with the 2018 New South Wales Physics Stage 6 Syllabus. The Student Book includes the very latest developments and applications of physics and incorporates best-practice literacy and instructional design to ensure the content and concepts are fully accessible to all students.

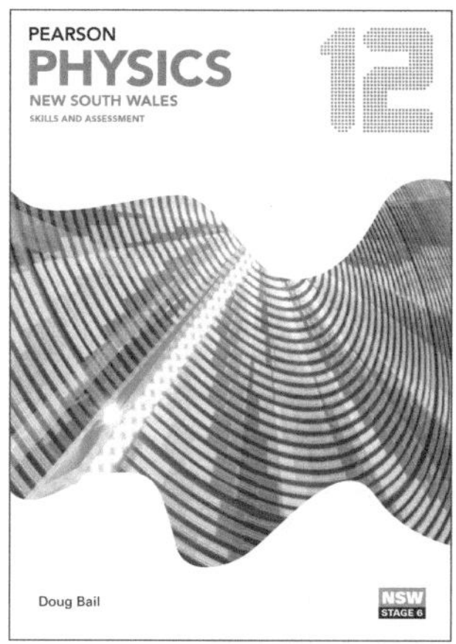

Skills and Assessment Book

The Skills and Assessment Book gives students the edge in preparing for all forms of assessment. Key features include a toolkit, key knowledge summaries, worksheets, practical activities, suggested depth studies and module review questions. It provides guidance, assessment practice and opportunities to develop key skills.

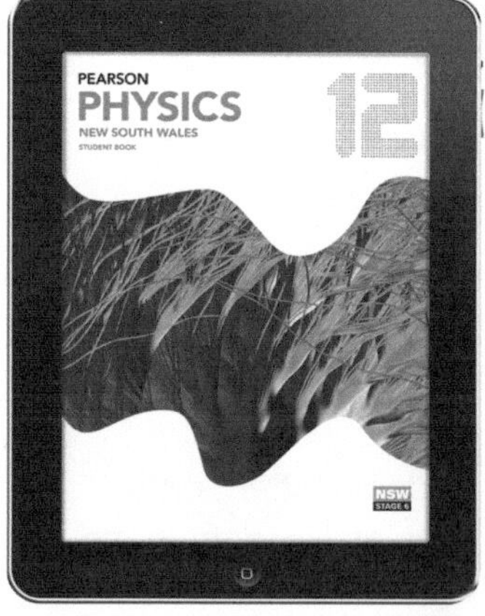

Reader+ the next generation eBook

Pearson Reader+ lets you use your Student Book online or offline on any device. Pearson Reader+ retains the look and integrity of the printed book. Practical activities, interactives and videos are available on Pearson Reader+ along with fully worked solutions to the Student Book questions.

Teacher Support

Online teacher support for the series includes syllabus grids, a scope and sequence plan, and three practice exams per year level. Fully worked solutions to all Student Book questions are provided, as well as teacher notes for the chapter inquiry tasks. Skills and Assessment book resources include solutions to all worksheets, practical activities, depth studies and module review questions; teacher notes, safety notes, risk assessments and lab technician's checklists and recipes for all practical activities; and assessment rubrics and exemplar answers for the depth studies.

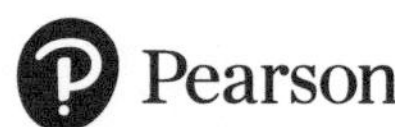

Access your digital resources at **pearsonplaces.com.au**
Browse and buy at **pearson.com.au**

ISBN 978 1 4886 1936 6

Physics toolkit

This toolkit provides support for developing the skills to undertake research and practical investigations, and complete depth studies. It also covers examination techniques and study skills. The toolkit can serve as a reference tool to be consulted as needed throughout the year.

Practical investigation

A practical investigation or activity is a laboratory experiment and may involve fieldwork. Experiments are conducted to test a hypothesis by generating primary data that is analysed to draw a conclusion. You are required to complete a minimum of 35 hours of practical work during this course, which may include a depth study practical investigation.

CONDUCTING A PRACTICAL INVESTIGATION

Practical investigations follow a precise scientific procedure. The following checklist provides a summary of essential elements of a science practical investigation and the experiment report. Refer to the checklist as you conduct a student-designed practical investigation.

Elements	Explanation	Tick ✔
Date	The date the practical was conducted and the report was written.	
Title/heading	A statement of what is being investigated.	
Purpose/aim	• A statement that describes what the scientist wants to investigate, demonstrate or find out in the experiment. • May be a statement or question. • May be one or two sentences. • Often written as 'To investigate the effect of *x* on *y*.'	
Hypothesis	• An educated guess about the expected result of the experiment. • A testable prediction, often in the form of 'If *x* is true and this is tested, then *y* will occur.'	
Variables	• Experiments measure relationships between variables and measure results. • Independent variable is the one that is changed. • Dependent variable is the one responding to the change of the independent variable. • Controlled variables are kept constant throughout the experiment.	
Risk assessments/ safety	• An analysis of the potential dangers of conducting the experiment. • Objective is to maximise safety by managing and minimising risk. • The risk assessment form on page xi guides you through the analysis of risk.	
Materials	• List of all equipment, chemicals and materials used. • Includes quantities (chemicals) and sizes (equipment).	
Procedure/method	• Detailed, numbered, step-by-step list of exactly what was done. • Includes quantities of chemicals used. • Can include 2D scientific drawings.	
Results	• A record of all observations and measurements taken during experiment. • Observations may be recorded as text, diagrams, photos or videos. • Most common records are presented as tables or graphs and can include calculations. • Check for errors and mistakes in data collected. Take measures to reduce these.	
Discussion	• Interpretation of results. • Includes analysis of the data, looking for trends and relationships. • Look for any problems such as bias, inaccuracy, unreliable data and errors. • Evaluate the method used, including a discussion of limitations. • Comment on whether results support or do not support the hypothesis.	
Conclusion	• Short statement based on the evidence that summarises the findings of the experiment. • Conclusion relates back to the purpose and the hypothesis, either supporting or not supporting the hypothesis. • Evidence should be clearly listed in support of the conclusion reached.	

RISK ASSESSMENT FORM

Five levels of safety should be considered in an investigation. The inverted pyramid ranks these levels in order of importance. The school and your teacher are responsible for reducing most of these risks. You as a student scientist can take measures to reduce the risks shown at the bottom of the hierarchy.

Complete the risk assessment form to identify possible risks for which you can take responsibility and to think of ways you can reduce risks to create a safe experiment environment.

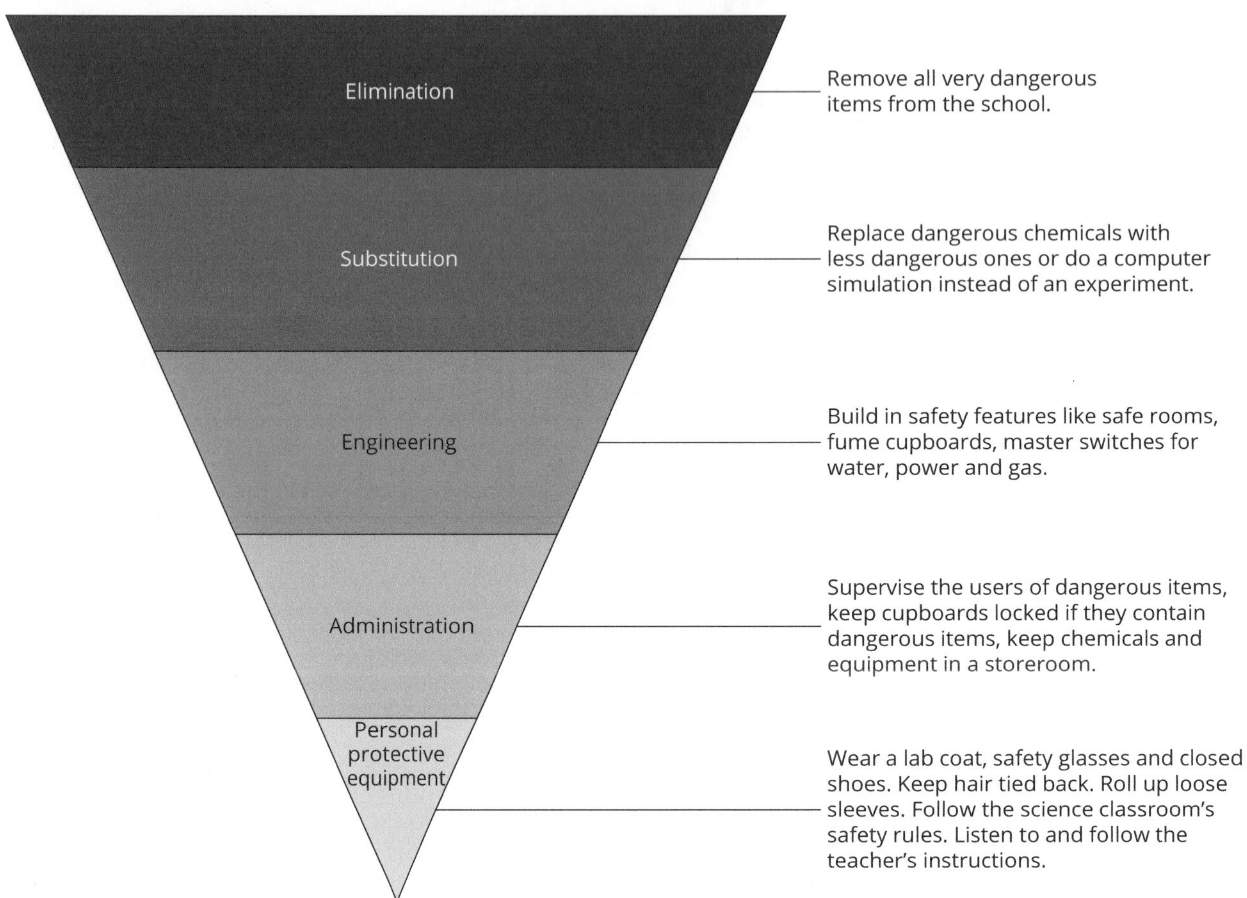

EXAMPLES OF PRACTICAL REPORTS

It can be difficult to gauge whether you have attained a high standard in your completed practical investigation report. Looking at sample practical reports can help you identify what is required. Two sample practical reports were provided in the Year 11 Skills and Assessment Book. They included annotations to draw your attention to key points to note on each practical report. These points are also reflected in the checklist, so you are able to use this as a tool to evaluate whether all requirements of the practical are complete. GO TO ➤ Year 11 Toolkit

 ISBN 978 1 4886 1936 6

Risk assessment form

What activity are you doing?

____

Title or description of the investigation

____

List:	**Identify any risks**	**State how you will:**
equipment you will be using		safely use each piece
chemicals or radioactive sources you will be using		carefully use each chemical or source carefully dispose of the chemicals
ethical issues you need to consider		ethically use animals in the laboratory ethically use human participants in the investigation
outdoor or fieldwork activities		reduce these risks
any other possible risks		reduce these risks

Secondary-sourced investigation

This section guides you in conducting a secondary-sourced investigation. For assistance with conducting a practical investigation, see pages ix–xi.

A secondary-sourced investigation is also often known as a research project. Such investigations require that you think carefully about the topic; find, collect and organise information; analyse and synthesise findings; and present your ideas. The investigation process is summarised in the following flow chart. An investigation is not necessarily a straightforward linear process as shown in the flow chart. You can move back and forth between steps as needed.

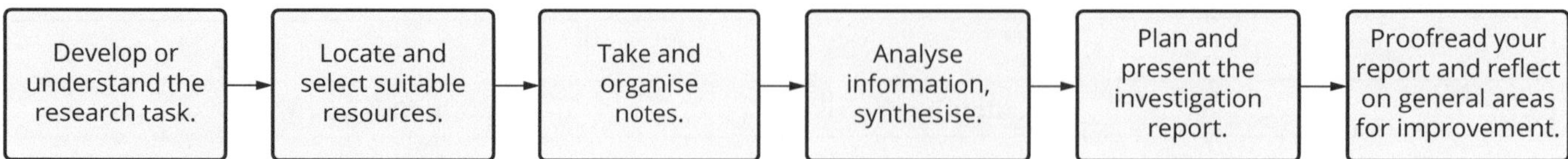

THE INVESTIGATION TASK

It is important to understand the breakdown of investigation questions or tasks. This helps you to write your own tasks and better understand the requirements of those written by others.

As you develop an investigation task, be aware of the depth of thinking it will require. The following chart provides support in writing tasks at different levels of thinking or complexity. It also provides some key words to help target tasks at these thinking levels and gives some examples of questions.

When developing your question or task, be conscious of the level at which you are pitching it. When you are writing your own questions for a depth study investigation, they should generally be at the analysis level.

Level of question complexity	Type of thinking	Words that may be used		Examples of questions
Simple	Retrieval of information—remembering, producing information on demand	• who • where • list • show • describe • select • complete • define	• what • when • label • demonstrate • name • state • recognise	**1** Define the term 'momentum'. **2** What factors affect inertia in a collision? **3** List the vector quantities in straight-line motion. **4** What variables affect the movement of thermal energy between two objects?
	Comprehension—the ability to understand information	• who • where • explain • represent • show how • represent	• what • when • summarise • draw • describe	**1** Represent the process of energy transfer with a well-labelled diagram. **2** Why are houses wired as a parallel circuit instead of a series circuit? **3** Explain why heat conductors feel cold to the touch. **4** What happens to a metal rod when it is heated?
	Analysis of information—scrutinising and breaking something into its smaller parts, including: • comparing • classifying • identifying errors • concluding • predicting • judging	• why • categorise • contrast • sort • organise • generalise • evaluate • edit • assess • judge	• how • compare • distinguish • discriminate between • deduce • critique • diagnose • identify errors • identify misunderstandings	**1** Organise the following in order from the simplest measure to highest level: velocity, acceleration, displacement. **2** Compare a solenoid with an electromagnet. **3** Compare and contrast alpha and beta decay.
Complex (requiring more thinking)	Application of information—using knowledge in new situations, including: • testing a hypothesis • solving a problem • experimenting and using data • decision-making	• why • investigate • find out about • test • solve • develop • decide	• how • research • experiment • predict • adapt • judge	**1** Early scientists believed that heat energy was transferred by the spilling of 'caloric'. Construct an argument for or against this statement. **2** Investigate the behaviour of a silicon diode in a simple circuit. **3** How would conduction of heat energy in an iron rod differ at different atmospheric temperatures?

 ISBN 978 1 4886 1936 6

RESOURCES

The resources you refer to in an investigation may be primary or secondary, or both.

Primary sources of information are from investigations that you have conducted yourself. Examples of primary sources are results from your experiments, reports of your scientific investigations, photographs you have taken, and specimens or artefacts that you have collected. Secondary sources of information are from investigations that have been conducted by others. Secondary sources include peer-reviewed articles, textbooks, biographies, documentaries, newspaper articles and many websites.

Refer to the two checklists below. The first shows features to look for when assessing and selecting the best sources for your investigation. The second shows the information that is required for the references section of your report.

Always remember to note the reference information for every resource you use. It is very time-consuming and difficult to backtrack to obtain these details when you need them to write the references.

Selecting resources for the investigation	Tick ✔
The resource is:	
• **credible** and I can identify the author, author's expertise and publisher	
• **current** because the date of publication of the material is provided and is recent	
• **factual** and I know that it is objective material and not biased	
• **accurate** and all information is correct	
• **relevant** and covers the area I am investigating	
• **readable** and neither too simple nor too complex in its coverage of the material	

Information for references and examples
Article in scientific journal or magazine Author(s), initials (year). Title of article. *Journal title, volume number*(issue number). page numbers. Kensrud, J.R., Nathan A.M. and Smith L.V. (2017). Oblique collisions of baseballs and softballs with a bat. *American Journal of Physics 85*(8). 503–509.
Book Author(s), initials (year). *Title of book* (edition, if not first). City: publisher. Rickard, G. *et al.* (2017), *Pearson Science Student Book 9*. Melbourne: Pearson Education. If there are more than three authors, list first author then write et al. (in italics), meaning 'and others'.
Internet Author(s) or name of organisation (year). *Title of web page or web document*. Retrieved from <URL>. American Institute of Physics (2017). *Periodic table*. Retrieved from http://history.aip.org/history/exhibits/electron/ Date accessed

NOTE-TAKING AND ORGANISING NOTES

Note-taking and organising notes takes skill. Good note-taking helps you avoid plagiarism and provides excellent information for writing up the investigation. Plagiarism is taking someone else's ideas and words and presenting them as your own work. You plagiarise if you copy sections or sentences from sources, or if you cut and paste from the internet. It is acceptable to use the ideas of others, but you must state clearly where the information has come from in the body of your report and include the source in your references.

The examples in the following table show the original text, plagiarised text, and acceptably rephrased text.

Original text	Plagiarised text	Rephrased text
Dogs have sweat glands on their feet. Dogs pant when they are hot because their sweat glands are not sufficient to cool down their bodies. In addition, their tongues allow the water from their bodies to evaporate and cool down their bodies.	Dogs have their sweat glands on their feet. When dogs get hot they pant because their sweat glands are not enough to cool down their bodies. Their tongues let the water from their bodies evaporate and cool their bodies.	Dogs pant in order to cool down. Water evaporates from their tongues and this lowers their body temperature. They can't get cool enough through just the sweat glands on their feet.

There are various approaches to effective note-taking. Whatever technique you use, try to keep notes brief, and focus on key points. Some examples include:

- dot point summary
- underlining or highlighting text
- labelled diagram
- flow chart: show sequences

- concept map: show connections between items
- Venn diagrams: show similarities and differences
- table: may incorporate any of the other note-taking techniques. Tables are useful for summarising longer and more complex information that has subparts. Adapt the table to suit your style and the task. The following sample table (which is partially completed) shows how this technique can be used to take notes for your secondary-sourced investigation.

Secondary-sourced investigation task Many scientists believe that limiting human population growth is necessary to control environmental damage. Construct an argument for or against this statement.			
	Source 1 (e.g. book) Title: Author: Publisher's name: Publisher's location: Date of publication:	**Source 2** (e.g. internet) Title: Author: URL: Date accessed:	**Source 3** (e.g. science journal) Author: Date: Title of article: Journal title: Volume number: Pages:
Population growth trends		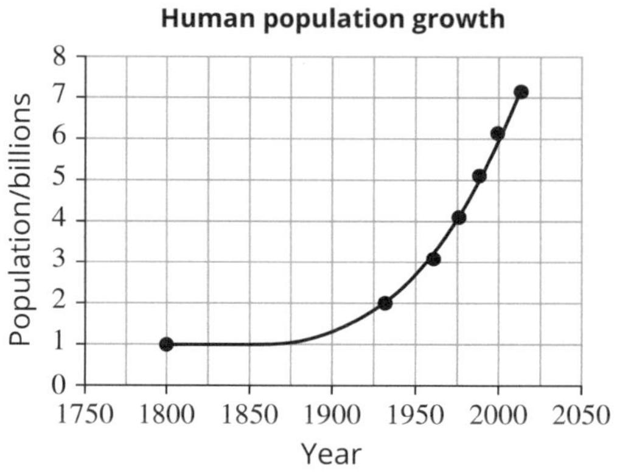 **Human population growth** Population/billions: 0–8 Year: 1750 1800 1850 1900 1950 2000 2050	
Impact of population growth on environment	• growth of cities • demand for resources		
Reasons to control and limit population growth	• increased demand for resources for human survival		
Reasons not to limit population growth but allow natural population growth			• other factors contribute to environment damage; land-use policies, e.g. poor land use

SCIENTIFIC WRITING

Scientists have a particular writing style. Your investigation should use this distinctive style to communicate your ideas. Writing in the scientific style is:

- objective—describes events rather than what people think or feel
- free of bias or personal opinion
- precise—avoids exaggeration and uses qualified language
- formal—uses scholarly language rather than colloquial or everyday language
- concise—conveys information in short, clearly understandable sentences without unnecessary information
- simple—uses short sentences where possible
- usually written in passive voice rather than active voice, although the active voice can be used sometimes
- structured to include headings, tables, diagrams and mathematical calculations.

ISBN 978 1 4886 1936 6

Examples of unscientific and scientific writing are demonstrated in the table below.

Unscientific writing	Scientific writing
Subjective, biased writing • The results were fantastic. • This produced a disgusting odour. • The breathtakingly beautiful bowerbird ...	**Objective, unbiased writing** • The results showed ... • This produced a pungent odour ... • The golden bowerbird ...
Exaggerated writing • The object weighed a huge amount. • The magnesium burst into huge flames. • Millions of ants swarmed over ...	**Accurate, precise writing** • The mass of the object was 250 kg. • The magnesium burnt vigorously. • Ants swarmed all over ...
Everyday, informal language • The bacteria passed away. • The results don't ... • We guessed that ... • Previous researchers were slack and missed ...	**Formal language** • The bacteria died. • The results do not ... • It was hypothesised that ... • Previous researchers did not perceive that ...
Active voice • We recorded oxygen levels every hour. • We put 50 g of solute in a conical flask containing distilled water, and then we slowly added 1 mol L^{-1} hydrochloric acid.	**Passive voice** • The oxygen level was recorded hourly. • 50 g of solute was placed in a conical flask containing distilled water, and then 1 mol L^{-1} of hydrochloric acid was slowly added.

PRESENTING THE INVESTIGATION

Scientific findings may be presented in a variety of ways. A common presentation format at science conferences is a poster. Posters can get ideas across to a large audience in an organised, concise and creative way. Other common presentation formats are essays, reports, orals or articles. Each presentation format has its own conventions. The table below summarises characteristics of a number of presentation formats.

Presentation formats and their characteristics		
Format	**Characteristics/inclusions**	
Poster	• balance of text and visuals • title, subheadings • balanced layout • captions for figures and tables	• references • hierarchy of font size according to subheading level • consistent font style: no more than three fonts
Report/article	• structured with introduction, paragraphs, conclusion • includes subheadings	• mainly text • can include diagrams, graphs, tables
Essay	• structured with an introduction, paragraphs, conclusion • introduction states focus of essay • each paragraph makes a new point supported by evidence	• each paragraph links back to last paragraph • a text-style presentation format—visuals at end in appendix • conclusion draws all ideas together but does not include any new information
Oral	• needs to be engaging • use cue cards but do not read from them • watch audience as you speak	• stand still and don't fidget • look at audience and appear confident

PROOFREADING

After you have completed the investigation and prepared your presentation, it is important to think about and check what you have done.

Proofread your work to minimise errors and maximise communication of the ideas from your investigation. Use these questions as a proofreading checklist.

Proofreading checklist	Tick ✓
Have I investigated the question fully?	
Have I expressed myself clearly so I communicated my ideas well?	
Have I used the scientific writing style?	
Have I included data analysis?	
Have I checked spelling, punctuation and grammar?	
Have I included references?	
Have I met the requirements of the presentation format?	

Depth study

A depth study is an investigation that allows you to look in more detail into a particular area of interest in the syllabus. The depth study gives you the chance to gain greater understanding of concepts and is designed so you take more responsibility for your own learning.

It is expected that you demonstrate the use of a variety of scientific skills. Depth studies may vary, and may include practical activities, field work reports, research assignments, and may be based on primary or secondary data or sources. To fulfil the minimum 15-hour time requirement, your teacher may ask you to complete one very detailed depth study or a number of smaller depth studies.

Your teacher may provide a scaffolded approach with questions that guide you through a depth study, or you may be asked to design it yourself. This section of the toolkit will help you with planning your own depth study.

PLANNING THE DEPTH STUDY

Careful planning of the depth study before actually starting the investigation will help you to be clear about what you will do and how you will do it. Think of yourself as taking on the role of your teacher or a scientist: identifying the depth study investigation topic, deciding how it will be investigated and presented, managing time through the investigation and checking that all syllabus requirements are met.

Clarify your ideas for your depth study, following these steps.

<table>
<tr><td>1 List all areas that interest you for your depth study. For example, forces in a collision or standing waves in a glass.</td><td>My areas of interest are:
____
____
____
____________</td></tr>
<tr><td>2 Do some quick research into each area of interest. Identify which gives you scope to develop into a depth study.</td><td>My chosen area of investigation is:
____
____________</td></tr>
<tr><td>3 Decide on the procedure you will use to conduct the depth study.
• A practical investigation:
- may begin with initial observations of a day-to-day activity
- may be a secondary-sourced investigation
- must include data analysis.
• A secondary-sourced investigation:
- may be expository (explain something)
- may be an argument based on evidence
- must include data analysis.
• Creating:
- may include design and construction of a working model
- may involve creating a portfolio.</td><td>My procedure will be:
____
____
____
____
____
____
____________</td></tr>
</table>

ISBN 978 1 4886 1936 6

4 Decide how you will present the depth study. Select a presentation format that suits your type of investigation. • A practical investigation - practical report • A secondary-sourced (research) investigation - documentary - media report - literature review - visual presentation - journal article - essay - scientific poster • Designing or inventing or creating • Portfolio • Working model	**I will present my depth study by:** ________________ ________________ ________________ ________________ ________________ ________________ ________________ ________________
5 Now that you have thought about what and how you will conduct the depth study, think about how you can best use your time. It is easy to spend most of the time conducting the practical or finding resources for the secondary-sourced investigation. This may leave little time to analyse data and prepare your presentation. Refer to the suggested planning timeline. Plan the depth study. 10% Conduct the investigation. 40% Organise information and analyse data. 30% Present information and proofread work. 20% Time allocation	**My depth study begins on** ________ **and is to be completed by** ________**.** Fill in the timeline to plan how you will allocate your time to complete the depth study, including specific dates.
6 Assessment requirements Depth study must: • address Working scientifically skills outcomes: Questioning and predicting, and Communicating • address a minimum of two additional skills outcomes: - Planning investigations - Conducting investigations - Processing data and information - Analysing data and information - Problem solving • must include at least one Knowledge and understanding outcome.	**My depth study will include these Working scientifically skills:** **1** Questioning and predicting **2** Communicating **3** ________________ **4** ________________ **My depth study covers this/these Knowledge and understanding outcome(s):** ________________ ________________

Study skills

There are a variety of techniques or strategies to help you study. You may find that you use different strategies in different situations. For example, you may prefer to highlight key phrases in your notebook throughout the year but make summaries of topics before an examination. The strategies you choose depend on personal preference and may not be the same as those used by classmates.

Effective study skills involve more than the learning strategies you use. Equally important is when you use skills. It is more effective to apply study skills throughout the year, revising and consolidating your knowledge as you progress through the course, rather than doing a rushed cram just before the examination. Revise work regularly. Being organised is the key to reducing your stress and setting up a study plan.

GETTING ORGANISED

To get yourself organised, try the following ideas.

- Use a diary to write down all homework and assessment tasks as soon as you get them. Note due dates and what you have to do.
- Be specific about the tasks you need to do. Rather than writing 'Do Physics' it is better to state things like which questions to answer and which page to look at in your student book.
- Write a list of everything you need to do each day. Tick off or cross out items as you complete them.
- Break down larger tasks into smaller separate parts that are manageable.
- Make sure your lists and planners are realistic. Do not set yourself more than you can actually do.

STUDY TECHNIQUES

Studying requires concentration. Remove any distractions and factor in some breaks. Have a break of 10 minutes every hour. Vary your study technique depending on the content to be learned and your personal preference. Although you may have already found a study technique that works for you, also consider the following options.

Study technique	Tips
Highlighting Radiation Radiation is a shortened form of electromagnetic radiation, which includes visible, ultraviolet and infrared light. Together with other forms of light, these make up the electromagnetic spectrum.The transfer of heat from one place to another without the movement of particles is by electromagnetic radiation. Electromagnetic radiation travels at the speed of light.	• Highlight or underline key points as you read your notes or text.
Summary notes *IONIC COMPOUNDS* • *Common properties of ionic compounds:* • *brittle, which means they shatter when hit with a hammer* • *hard, which means they are resistant to scratching* • *high melting point* • *unable to conduct electricity in solid state* • *good electrical conductor in liquid or aqueous states*	• Create a list of key headings and add some dot points about each heading. • Write your own summary of the key ideas in each chapter. • Use headings and subheadings. • Underline key words and key phrases. • Use simple diagrams. • The most effective chapter summaries are clear, to the point and uncluttered.
Diagrams 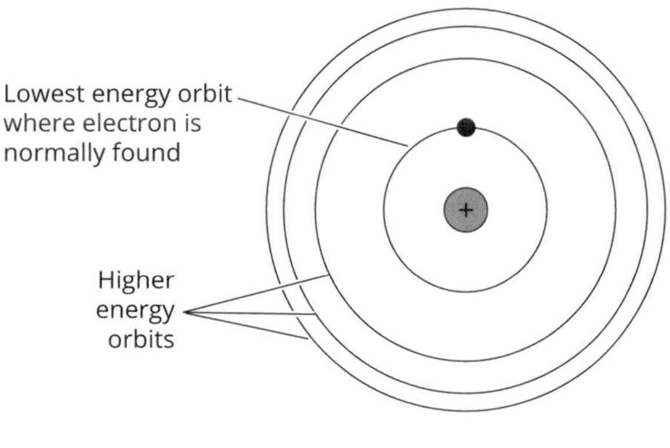 Lowest energy orbit where electron is normally found + Higher energy orbits	• Diagrams can be used as a summary of key concepts. • They are useful memory triggers. • Diagrams cover a lot of information in a visual way, with minimal text.
Concept maps 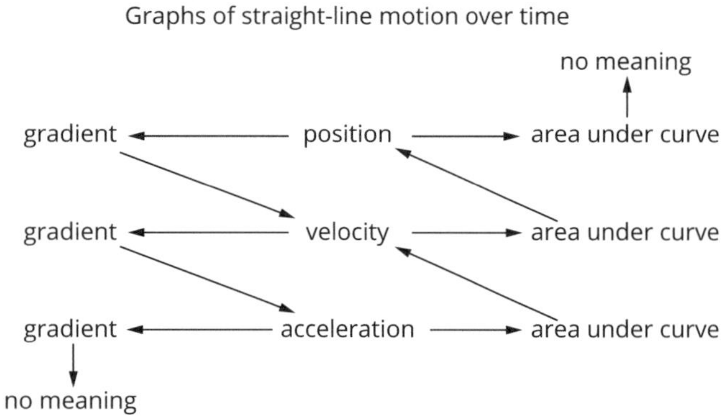 Graphs of straight-line motion over time no meaning gradient ← position → area under curve gradient ← velocity → area under curve gradient ← acceleration → area under curve no meaning	• These are a great way of connecting key terms and ideas in a simple and ordered way. • Use the lines that connect ideas to note the relationships between the words and phrases. • They can include text and images. • They can be a simple or more complex summary tool. • Concept maps and other graphic organisers such as Venn diagrams and flow charts show how information is connected and help deepen understanding.

 ISBN 978 1 4886 1936 6

<table>
<tr><td>Tables
<table>
<tr><td colspan="3">Subatomic particles</td></tr>
<tr><td>Relative mass</td><td>Charge</td><td>Location</td></tr>
<tr><td>1</td><td>neutral</td><td>nucleus</td></tr>
<tr><td>1</td><td>positive</td><td>nucleus</td></tr>
<tr><td>$\frac{1}{1836}$</td><td>negative</td><td>orbiting nucleus</td></tr>
</table></td>
<td>• Tables are useful to show relationships between different factors.
• Information is uncluttered.</td></tr>
<tr><td>Mnemonic devices
Work is fun, sir.
... triggers your memory for ...
$W = Fs$</td>
<td>• Mnemonic tools help you remember information.
• To remember that work is a product of force and displacement, memorise the sentence shown, which contains the initial letters of each variable.</td></tr>
<tr><td>Glossary
Electron: negatively charged particle in an atom
Diode: a semiconductor</td>
<td>• Compile your own glossary; writing the terms and definitions will make them easier to remember.
• Add images to help you remember.
• Write each term in a sentence.
• Memorise these terms and definitions.</td></tr>
<tr><td>Trigger words
Origin of universe– two theories:
• Big Bang
• Steady state theory – Fred Hoyle</td>
<td>• Write down the key words associated with a topic or theme.
• Trigger words are useful in helping to remember other related words and ideas.</td></tr>
<tr><td>Repeating information aloud</td>
<td>• This is a good way to remember and to force yourself to slow down and absorb the information.</td></tr>
<tr><td>Practising</td>
<td>• Do as many review questions and old exams as possible.</td></tr>
<tr><td>Flash cards
What is the Schrödinger model?
Identify the states of matter and changes between them.</td>
<td>• Making flash cards helps your understanding.
• Make your own cards.
• Write a question on one side and the answer on the back.
• Cards can include definitions, brief explanations, diagrams, equations, graphs.</td></tr>
<tr><td>Teaching someone</td>
<td>• Teach friends or family members.
• Teaching a difficult concept to someone means you must first understand the concept yourself.</td></tr>
<tr><td>Writing notes</td>
<td>• Hand-write rather than type summary notes.
• Remember the examination requires you to write answers.
• Practise writing for long stretches of time and make sure your writing is legible.</td></tr>
<tr><td>Responding to feedback and self-correcting</td>
<td>• Check through all feedback from your teacher.
• Highlight what was right or wrong.
• Attempt to identify where you have errors and rework the answer to get it right.</td></tr>
</table>

EXAMINATION PREPARATION

In the weeks before the examination begin your exam preparation. The earlier you begin revising, the easier it will be.

Like most skills, practice will improve your ability to do exams and to handle different types of exam questions. Doing practice exams is vital because you gain experience in:

- using reading time effectively before starting to write
- allocating the right amount of time for each question
- working to a time limit
- reading and interpreting questions
- understanding what is required of each question
- planning answers
- deciding on relevant information
- proofreading/checking over your own answers
- writing efficiently for the duration of the exam.

Use your practice exam experience not only to revise but also to analyse your strengths and weaknesses, and to assess how you managed your time.

Use the following checklist as a reminder of your study program.

Study program	Tick ✓
I have revised all areas of the course.	
I have highlighted important points.	
I have made a summary of the important points in each topic.	
I have read over my revision notes.	
I have looked at and worked through sample exam papers.	
I have answered practice questions in the appropriate time limit.	

EXAMINATION STRATEGIES

Familiarise yourself with the conditions of the examination well before the day you sit the exam. You should know:

- the number of exams for the subject
- the amount of time allowed for reading before the exam begins
- the amount of writing time allowed
- any particular equipment allowed and/or required, such as a calculator, lead pencils, pens, ruler
- strategies to tackle the exam. Exam strategies are listed in the table that follows.

Exam strategies

Reading time

- You will be given 5 minutes of reading time.
- No writing at all is allowed during this time—no note-taking, no highlighting, no underlining.
- Read the instructions.
- Read through the short-answer questions first—this will give you an overall sense of the themes of questions that require written responses.
- Read the multiple-choice questions next.
- Decide the order in which you will answer questions. Start with what you consider to be the easiest question to build confidence.

Writing time

- You will be given 3 hours of working time.
- Begin with the multiple-choice questions.
- Answer every multiple-choice question, even if you can only make an educated guess.
- If you are unsure of an answer to a multiple-choice question, mark it so you can come back to it if time allows.
- Attempt the short-answer questions next.
- Attempt the easiest short-answer questions first and work your way to the more challenging questions.

Tips for answering questions

- Carefully read each question, underlining key words.
- Be aware that most questions are structured so they become more challenging towards the end. You may not be able to answer the last part of a question but you can earn most of the marks by answering the easier parts of the question.
- Look carefully at any diagrams, pictures, tables and graphs and make sure you understand their relevance to the questions involved.
- For questions with graphs, read the graph title and the labels on the axes carefully, so that you can establish the relationship the graph is showing.
- For questions with tables, read the headings on the columns and rows carefully, so that you can analyse the content of the table effectively.
- Check for the key words in a question. Highlight them but don't colour the whole question.
- Plan your answers before you write, remembering to address the exam criteria.
- For questions with parts, read the whole question first. This gives you an overall picture of the question. It will also help to ensure that you do not repeat yourself in subsequent parts of the question.
- Make sure you actually answer the question that is asked.
- Once you have answered the question, re-read your answer and then re-read the question, to ensure that you have actually answered the question.
- When writing a definition, don't use the word you are defining in the actual definition.
- If giving values from a graph, use a ruler to line up points with the axes so you can be accurate, and always include units in your answer.
- Be sure to attempt all questions.
- Read over your answers to pick up careless errors—the mind is faster than the hand, and you may not always write what you intend (especially when you have limited time).
- Write legibly. Exams are scanned and marked online. If the assessor can't read your answer, they can't mark it as correct.
- Keep an eye on the time.
- Never leave an exam early. Use any spare time to re-read and check your answers.

Exam cues

- The number of marks allocated to a question provides a clue about how much you are expected to write. Two marks usually means you need to make a minimum of two points.
- The number of lines allowed for the answer indicates the length of the expected answer. If your writing is large, you may need to turn the page and continue on the back of the page. Make sure you indicate that the examiner must turn to the back of the page for the rest of the answer. Where the answer continues, clearly state that it is the continuation of the question and state the question number.

HANDLING EXAM QUESTIONS

The exam is divided into two sections: Section I consists of 20 multiple-choice questions and Section II contains a series of short-answer questions.

Multiple-choice questions	Short-answer questions
• In many cases you can use your basic knowledge of the subject to cancel out one or two options. • Mark an answer to every question on your answer sheet as you go through them. This avoids leaving any questions unanswered if you run out of time, and avoids getting out of order by leaving blank rows. You can always go back and change the questions you were unsure about. • Circle the letter of your answer to the question on the exam paper as well, so you can easily check if you get out of sync on the answer sheet. • If you are finding a question very difficult, don't spend too much time on it as it is only worth 1 mark. Guess an answer if necessary, then mark the question on your paper and come back to it at the end if you have time. • If you have time at the end of the exam, redo the questions with your original answer covered, and compare the answers.	• Be careful to do what the question asks. For example, do you need to list, describe, explain or compare? Each requires a different response. • When in doubt, point it out, i.e. give a thorough answer. Dot points and labelled diagrams are acceptable as long as you fully address the question. There is no need for long paragraphs in physics. Keep worded explanations short and to the point. • Tailor your answer specifically to the question being asked. Don't rely on a general answer when given a specific scenario. For example, indicate the direction of change, don't just say it changed. • Look at the marks allocated for an indication of the level of detail required. Generally, 1 mark = 1 key point, 2 marks = 2 key points, etc. • Set out your answers clearly, stating the formulae you intend to use as this may earn marks; for example, $\varepsilon = -N\frac{\Delta\Phi}{\Delta t}$. • Show your working—because it is easier for you to check and because you often gain part marks and consequential marks if the marker can see where you made an error. • Read over and use any values listed in the data booklet given to you with the exam paper, such as the mass of an electron. • Take care to include units in your answers if they are required. • Use an appropriate number of significant figures in your results. It is important to record data to the number of significant figures available from the equipment or observation. Using either a greater or smaller number of significant figures can be misleading and can lose you marks.

Every year examiners comment that many students fail to answer the question; that is, students' answers are 'not relevant' and are 'off the point'. Giving a relevant answer to a question is vital and depends on:

- choosing appropriate content
- using content purposefully to do what is asked (see the Key action word list below).

Key action word list	
Analyse	Break down into key points; show the essence of; inquire into; identify components and the relationship between them; draw out and relate implications
Compare	Show points of similarity and points of difference
Describe	Give the general features/characteristics of
Design	Provide steps for an experiment or procedure
Evaluate	Make a judgement based on criteria; determine the value of
Explain	Provide details to give the reader an understanding of; relate cause and effect; make the relationships between things evident; provide why and/or how
Justify	Demonstrate the correctness of (by giving evidence); give evidence in support of an argument or conclusion
Suggest	Put forward ideas, proposals, recommendations
How effective...?	Make a judgement based on criteria; determine the value of

EXAMPLES OF EXAM RESPONSES

Each year the examiners write an assessment report on the last exam. Reading these reports is a good way to understand what the examiners expected in the answers. The reports show where students showed strong responses and where improvement could be found.

Looking closely at sample responses will help you identify what is required to gain full marks. Two sample responses are provided below: one has room for improvement (low-level response) while the other is well done (high-level response). The annotations draw your attention to key points to note in each response.

QUESTION 1 (9 MARKS)

In Millikan's oil-drop experiment, charged drops were suspended within an electric field, as shown in the diagram below.

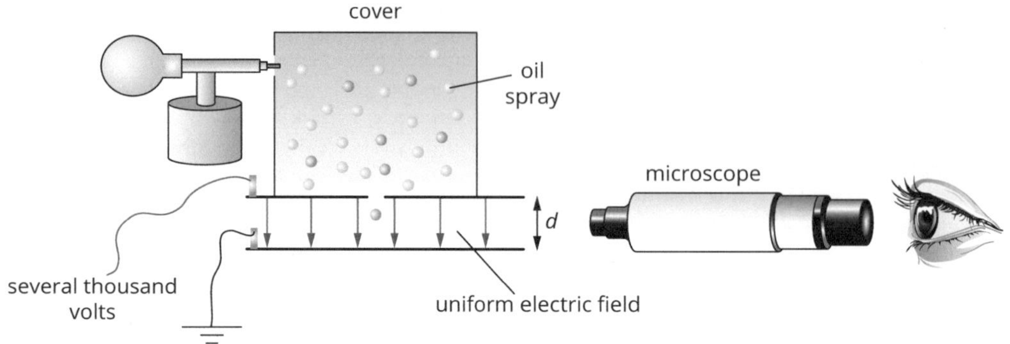

A group of scientists attempted to recreate Millikan's experiment by calculating the weight of different drops which become suspended between a potential difference when the electric field strength is increased. They created the following hypothesis and purpose:

> If a droplet of oil spray is successfully suspended within an electric potential, then at this point the weight force must be balanced by the electrostatic force felt by the particle. The electrostatic force is governed by the elementary charge q_e, which will be determined experimentally by finding the weight of suspended particles at different potential differences.

After completing their experiment, the following table was created:

Electric field strength ($\times 10^5$ V m^{-1})	Weight ($\times 10^{-14}$ N)
1.25	2.01
1.50	2.42
1.75	2.72
2.00	3.22
2.25	3.62

 ISBN 978 1 4886 1936 6

High-level responses

Question 1

a Plot the data and draw a line of best fit.

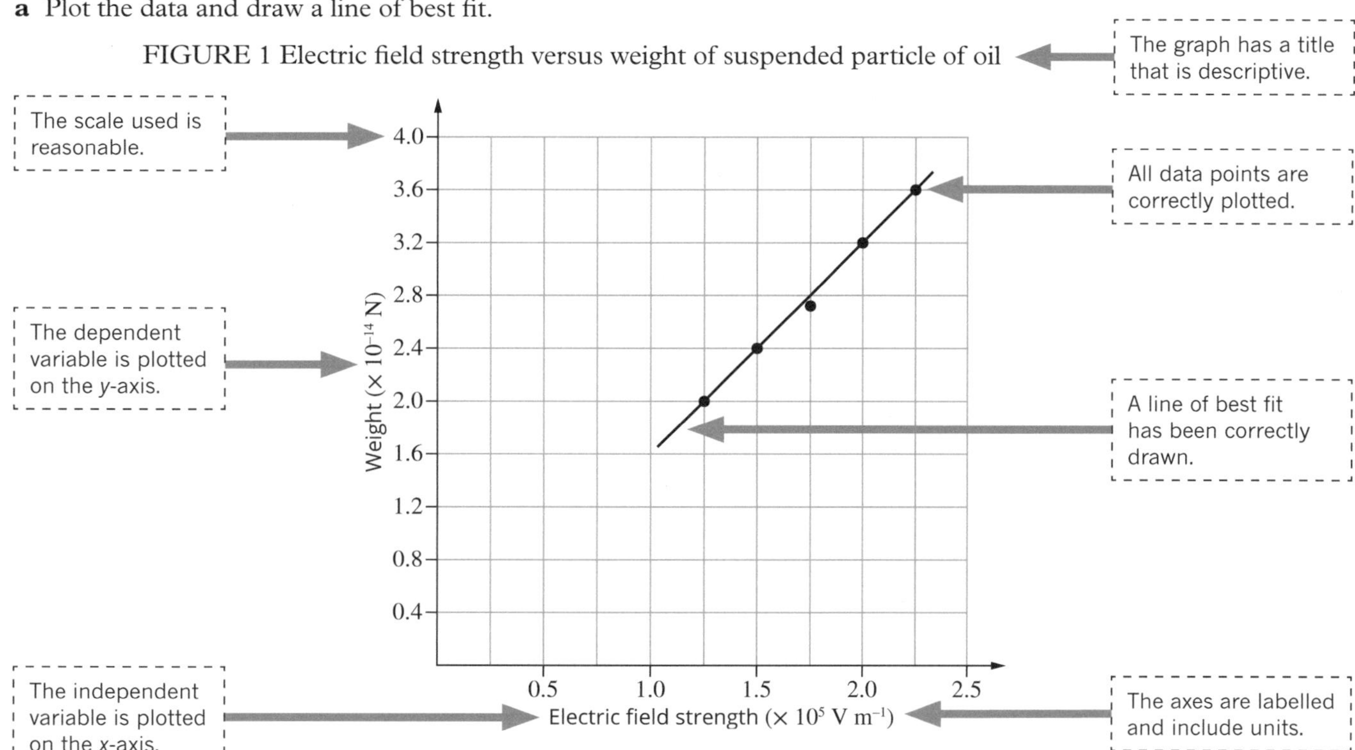

b Describe qualitatively the relationship between the electric field strength and the weight of the oil drop. In your response, explain what variable the gradient of the graph represents.

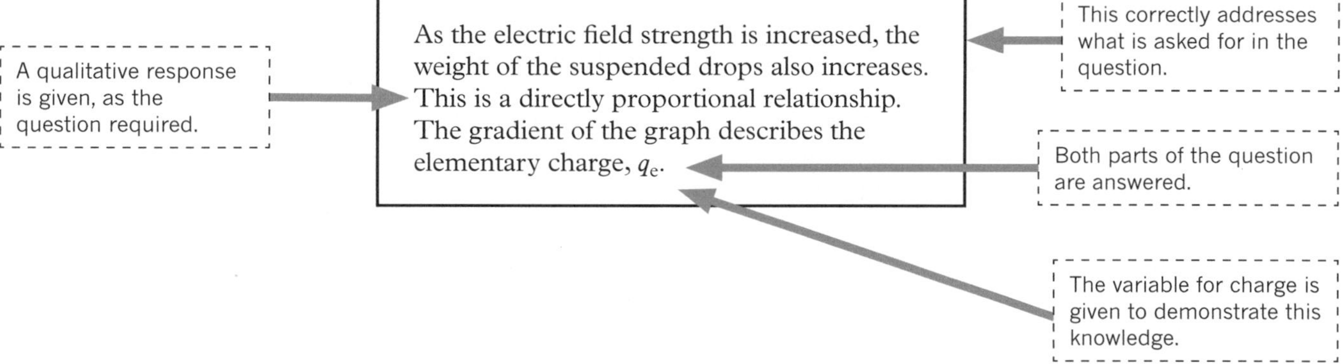

c Compare the value of the elementary charge determined by this experiment to the theoretical value of the charge of an electron.

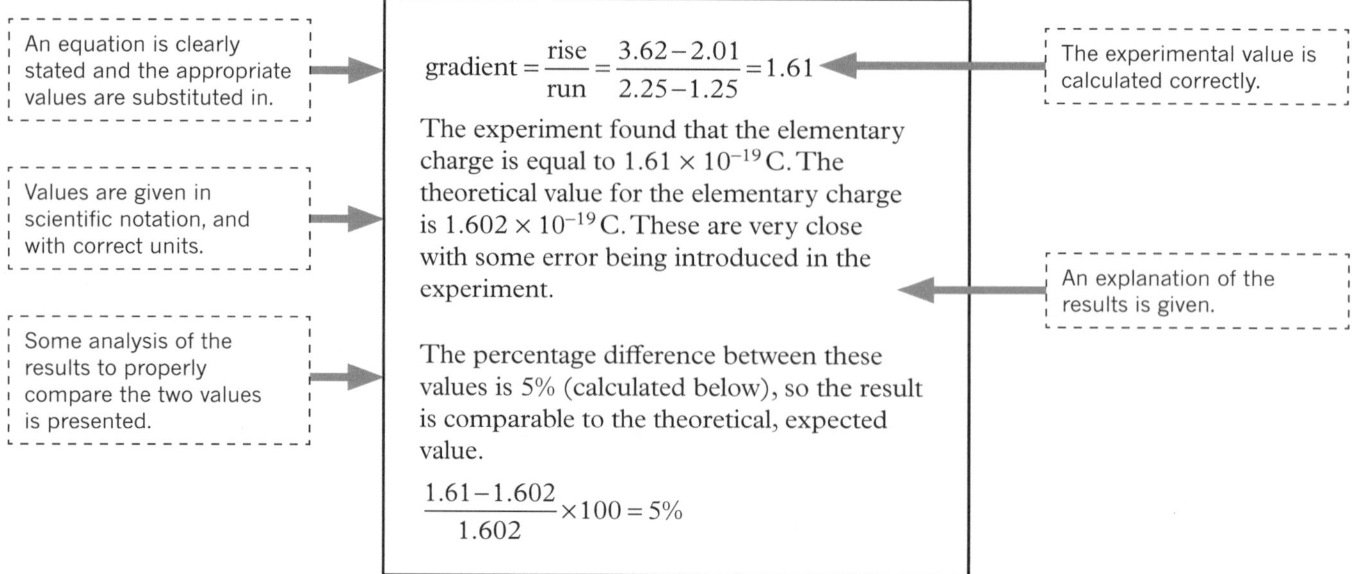

d Define the term validity and describe the validity of these results.

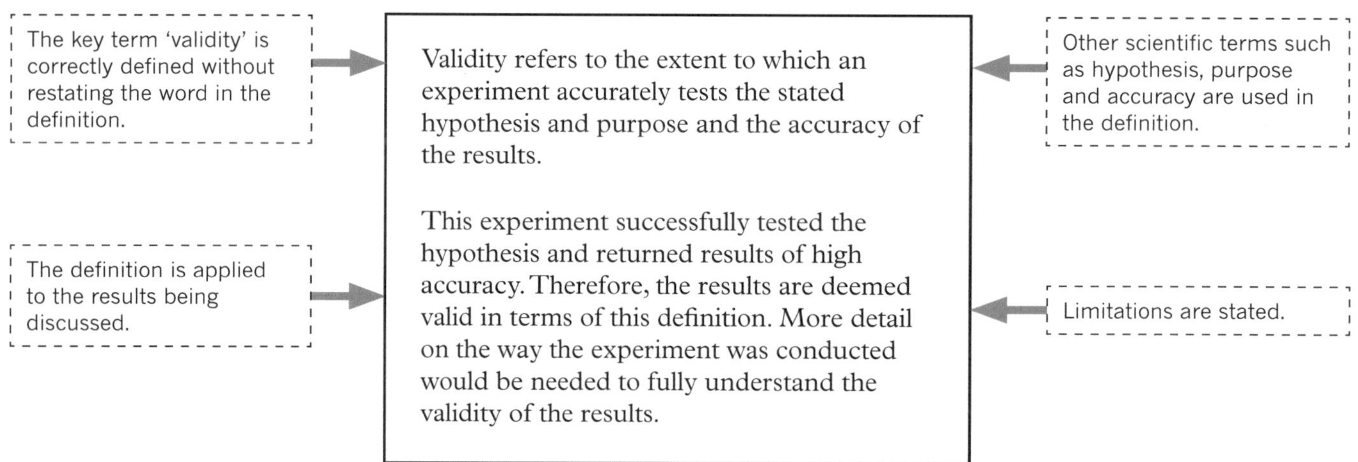

 ISBN 978 1 4886 1936 6

Low-level responses

Question 1

a Plot the data and draw a line of best fit.

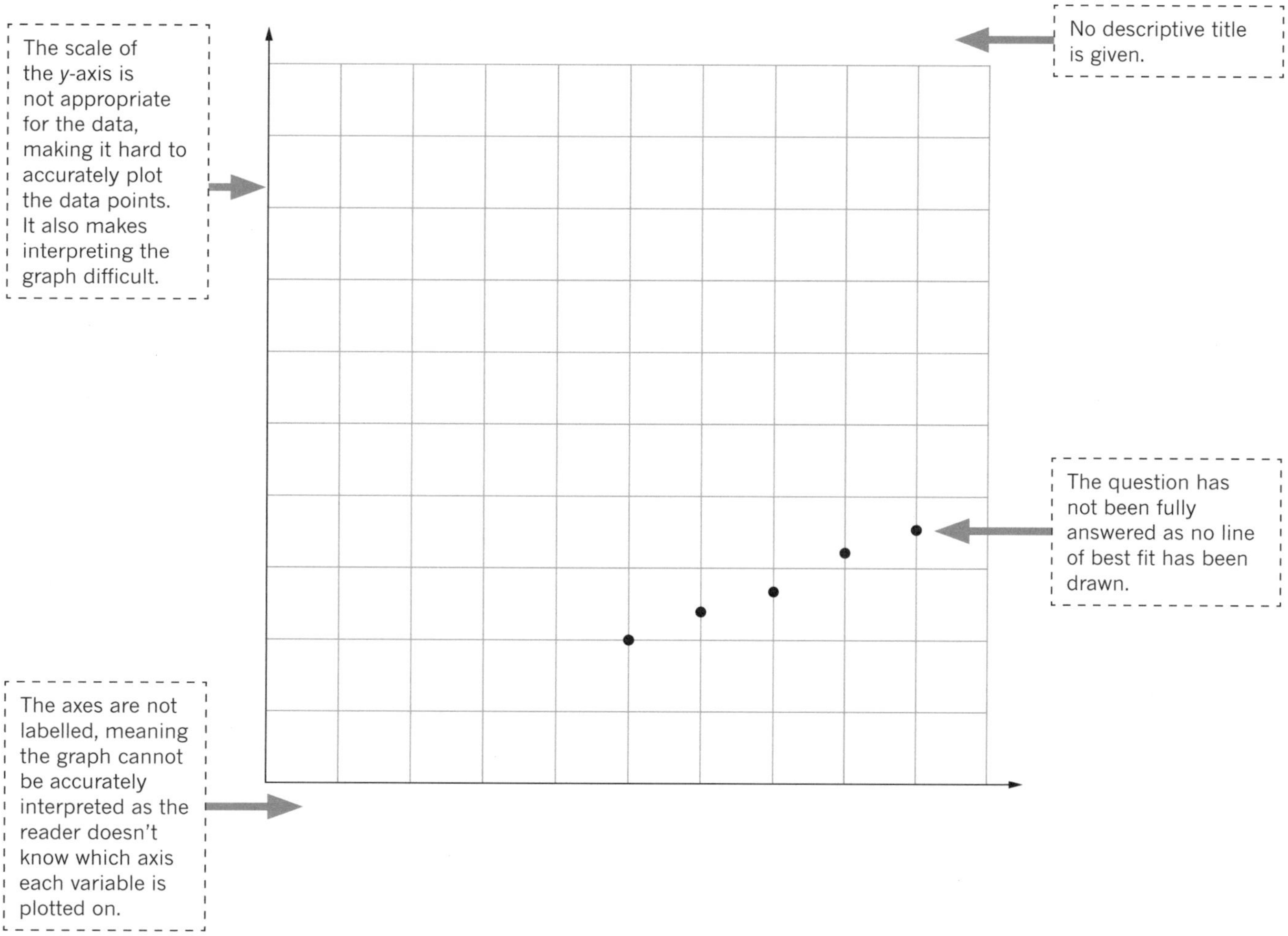
The scale of the *y*-axis is not appropriate for the data, making it hard to accurately plot the data points. It also makes interpreting the graph difficult.
No descriptive title is given.
The question has not been fully answered as no line of best fit has been drawn.
The axes are not labelled, meaning the graph cannot be accurately interpreted as the reader doesn't know which axis each variable is plotted on.

b Describe qualitatively the relationship between the electric field strength and the weight of the oil drop. In your response, explain what variable the gradient of the graph represents.

The equation given describes the quantitative relationship between the variables, but the question explicitly asked for a qualitative analysis.

The relationship between the electric field strength and the weight can be given by the following equation: $F_g = 1.61\,E$.

The gradient is given in the equation but the answer doesn't explain what it represents.

Neither parts of the question are answered satisfactorily.

ISBN 978 1 4886 1936 6

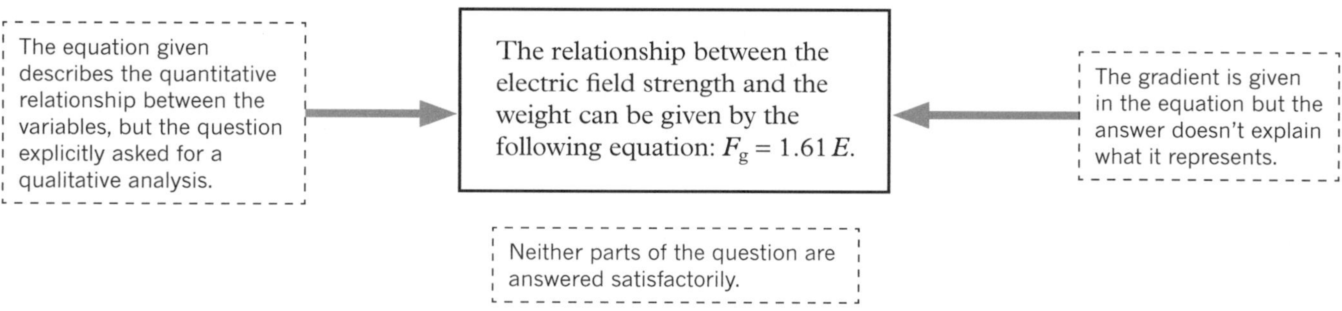

c Compare the value of the elementary charge determined by this experiment to the theoretical value of the charge of an electron.

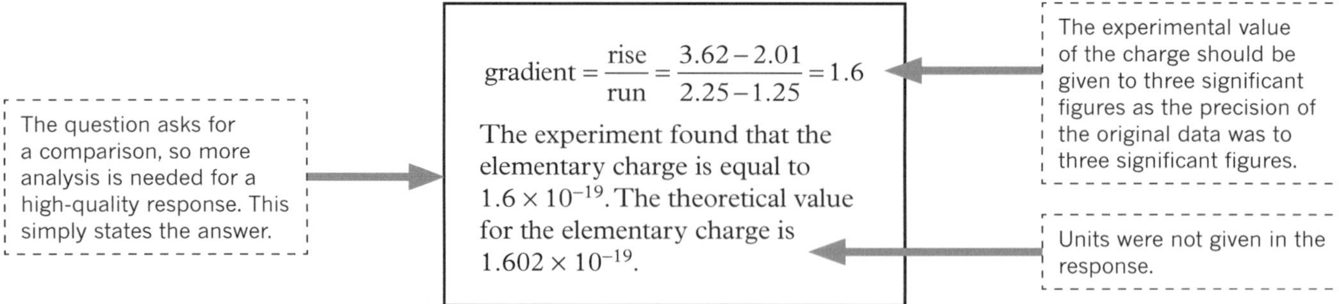

d Define the term validity and describe the validity of these results.

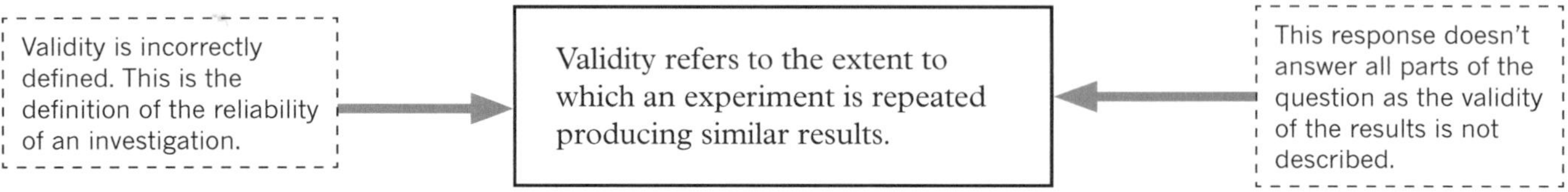

GETTING READY

Once you have finished revising and feel confident with the material, use the checklist below to make sure you are prepared before the exam day arrives.

The exam itself	**Tick ✔**
Have you:	
• checked the exact exam requirements?	
• obtained and carefully read the instruction cover sheet from previous exams? (Check that there are no changes expected for your exam.)	
• planned how to best use your reading time?	
Practical steps	
Have you:	
• assembled all materials required and allowed? This includes: - black pens to write your answers - pencils of the correct lead for the multiple-choice answer sheet - sharpeners, erasers, rulers and other permissible materials like calculators and spare batteries - a water bottle (check that your bottle meets the requirements—it may need to be clear with no label) - a watch to keep track of the time. (You are not permitted to look at mobile phones.)	
• double-checked dates, times and locations for each exam?	
• remembered your student number?	
• packed a form of photo ID (if required)?	
• arranged transport, allowing ample time to avoid rushing and to manage possible delays? (Be sure to arrive early to settle nerves and gather thoughts.)	
• planned for appropriate meals and drinks, as exam times vary? (Avoid heavy meals before exams as they make you sleepy.)	

 ISBN 978 1 4886 1936 6

MODULE 5 Advanced mechanics

Outcomes

By the end of this module you will be able to:

- select and process appropriate qualitative and quantitative data and information using a range of appropriate media PH12-4
- analyse and evaluate primary and secondary data and information PH12-5
- solve scientific problems using primary and secondary data, critical thinking skills and scientific processes PH12-6
- communicate scientific understanding using suitable language and terminology for a specific audience or purpose PH12-7
- describe and analyse qualitatively and quantitatively circular motion and motion in a gravitational field, in particular, the projectile motion of particles PH12-12.

Content

PROJECTILE MOTION

INQUIRY QUESTION **How can models that are used to explain projectile motion be used to analyse and make predictions?**

By the end of this module you will be able to:

- analyse the motion of projectiles by resolving the motion into horizontal and vertical components, making the following assumptions:
  - a constant vertical acceleration due to gravity
  - zero air resistance
- apply the modelling of projectile motion to quantitatively derive the relationships between the following variables:
  - initial velocity
  - launch angle
  - maximum height
  - time of flight
  - final velocity
  - launch height
  - horizontal range of the projectile (ACSPH099)
- conduct a practical investigation to collect primary data in order to validate the relationships derived above
- solve problems, create models and make quantitative predictions by applying the equations of motion relationships for uniformly accelerated and constant rectilinear motion. ICT N

Module 5 • Advanced mechanics

CIRCULAR MOTION

INQUIRY QUESTION **Why do objects move in circles?**

By the end of this module you will be able to:

- conduct investigations to explain and evaluate, for objects executing uniform circular motion, the relationships that exist between:
    - centripetal force
    - mass
    - speed
    - radius
- analyse the forces acting on an object executing uniform circular motion in a variety of situations, for example:
    - cars moving around horizontal circular bends
    - a mass on a string
    - objects on banked tracks (ACSPH100) CCT ICT
- solve problems, model and make quantitative predictions about objects executing uniform circular motion in a variety of situations, using the following relationships:
    - $a_c = \frac{v^2}{r}$
    - $v = \frac{2\pi r}{T}$
    - $F_c = \frac{mv^2}{r}$ ICT N
    - $\omega = \frac{\Delta\theta}{t}$
- investigate the relationship between the total energy and work done on an object executing uniform circular motion
- investigate the relationship between the rotation of mechanical systems and the applied torque
    - $\tau = r_{\perp}F = rF\sin\theta$ ICT N

MOTION IN GRAVITATIONAL FIELDS

INQUIRY QUESTION **How does the force of gravity determine the motion of planets and satellites?**

By the end of this module you will be able to:

- apply qualitatively and quantitatively Newton's law of universal gravitation to:
    - determine the force of gravity between two objects $F = \frac{GMm}{r^2}$
    - investigate the factors that affect the gravitational field strength $g = \frac{GM}{r^2}$
    - predict the gravitational field strength at any point in a gravitational field, including at the surface of a planet (ACSPH094, ACSPH095, ACSPH097)

- investigate the orbital motion of planets and artificial satellites when applying the relationships between the following quantities: CCT ICT N
  - gravitational force
  - centripetal force
  - centripetal acceleration
  - mass
  - orbital radius
  - orbital velocity
  - orbital period
- predict quantitatively the orbital properties of planets and satellites in a variety of situations, including near the Earth and geostationary orbits, and relate these to their uses (ACSPH101) ICT N
- investigate the relationship of Kepler's laws of planetary motion to the forces acting on, and the total energy of, planets in circular and non-circular orbits using: (ACSPH101)
  - $v = \frac{2\pi r}{T}$
  - $\frac{r^3}{T^2} = \frac{GM}{4\pi^2}$ ICT N
- derive quantitatively and apply the concepts of gravitational force and gravitational potential energy in radial gravitational fields to a variety of situations, including but not limited to: ICT N
  - the concept of escape velocity $v_{esc} = \sqrt{\frac{2GM}{r}}$
  - total potential energy of a planet or satellite in its orbit $U = -\frac{GMm}{r}$
  - total energy of a planet or satellite in its orbit $U + K = -\frac{GMm}{2r}$
  - energy changes that occur when satellites move between orbits (ACSPH096)
  - Kepler's laws of planetary motion (ACSPH101).

Key knowledge

Vectors can be represented using a variety of different notations. Throughout this book you will see some vectors written with an arrow above the variable, i.e. $\vec{F}$. This is known as vector notation. There are different sorts of vector notation; for example, the same vector can be represented as $\tilde{F}$, $\boldsymbol{F}$ or F. Regardless of the way a vector is represented, it is important for you to understand that the same rules apply when completing any calculation involving vectors. A vector is a variable which has both a magnitude and a direction. Being able to understand when a variable is described by a vector is an important skill for a physicist to develop.

Projectile motion

A **projectile** is any object that is thrown or projected into the air and has no power source driving it. If air resistance is ignored, the only force acting on a projectile is its **weight**, i.e. the force of gravity, F_g. This force can be assumed to be constant and is always directed vertically downwards with an **acceleration** of 9.8 m s^{-2}. Projectiles move in parabolic paths that can be analysed by considering the horizontal and vertical components of their motion (shown in Figure 5.1). The two factors that affect a projectile's motion are the angle at which it is launched and its initial **velocity**.

FIGURE 5.1 The motorcycle and rider are travelling in a parabolic path as they fly through the air.

PROJECTILE LAUNCHED HORIZONTALLY

Projectiles can be launched at any angle. The launch velocity needs to be resolved into vertical and horizontal components using trigonometry to complete most problems. For projectiles launched horizontally, the horizontal velocity is constant if air resistance is ignored and is equal to the horizontal launch velocity; the initial vertical velocity is zero and will increase during the flight. A parabolic path is traced by an object accelerating only in the vertical direction while moving at constant velocity in the horizontal direction. The following equations of motion for uniform acceleration must be used for the vertical component of the motion:

> $v = u + at$
>
> $s = ut + \frac{1}{2}at^2$
>
> $v^2 = u^2 + 2as$
>
> where:
>
> s is the **displacement** (m)
>
> u is the initial velocity (m s^{-1})
>
> v is the final velocity (m s^{-1})
>
> a is the acceleration (m s^{-2})
>
> t is the time (s).

As the horizontal velocity of a projectile remains constant throughout its flight, the following equation for average velocity can be used for this component of the motion:

$$v_{av} = \frac{s}{t}$$

To solve a projectile motion problem, a diagram should be constructed where the direction convention is clearly specified (whether up or down is the positive or negative direction). The information supplied for the horizontal and vertical components should be drawn separately.

For example, a ball is hit horizontally from the top of a 10.5 m cliff with a speed of 5.0 m s^{-1} (Figure 5.2). Use acceleration due to gravity of 9.80 m s^{-2} and ignore air resistance. To calculate the time it takes the ball to land and the distance it has travelled from the base of the cliff, there are a few steps you should take.

By first writing out the variables that are known, you can then decide on the equations of motion that are best suited to finding the answers. Remember to keep the horizontal and vertical components of motion separate.

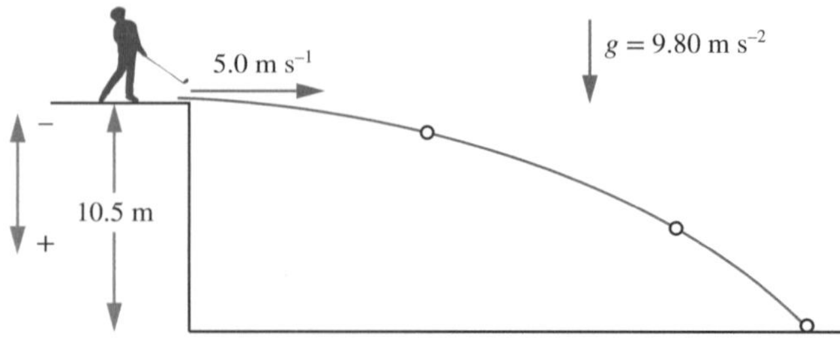

FIGURE 5.2 Parabolic motion with initial horizontal velocity only

The time can then be found using the vertical components of motion and the equation: $s = ut + \frac{1}{2}at^2$.

 ISBN 978 1 4886 1936 6

$$10.5 = 0 + \tfrac{1}{2} \times 9.80 \times t^2$$

$$t = \sqrt{\frac{10.5}{4.9}}$$

$$= 1.46\,\text{s}$$

Similarly, to calculate the horizontal displacement (i.e. how far the ball travelled from the base of the cliff), you can use the equation: $v_{av} = \frac{s}{t}$.

$$5.0 = \frac{s_H}{t}$$

$$s = 5.0 \times 1.46$$

$$= 7.3\,\text{m}$$

Generally, vector answers need both their magnitude and direction, so pay close attention to how a question is written to know what information is needed in your solution. In the example given above, you were asked for the distance travelled, not the horizontal displacement, so the answer is given as a scalar.

PROJECTILES LAUNCHED OBLIQUELY

The previous section looked at projectiles that were launched horizontally. A more common situation is projectiles that are launched obliquely (at an angle), by being thrown forwards and upwards at the same time (like in Figure 5.3). For objects initially launched at an angle to the horizontal, it is useful to calculate the initial horizontal and vertical velocities using trigonometry. The initial velocity can be broken down into its perpendicular horizontal (u_H) and vertical (u_V) components using the trigonometric relationships below:

$$\sin\theta = \frac{\text{opposite}}{\text{hypotenuse}} = \frac{u_V}{u} \qquad u_V = u\sin\theta$$

$$\cos\theta = \frac{\text{adjacent}}{\text{hypotenuse}} = \frac{u_H}{u} \qquad u_H = u\cos\theta$$

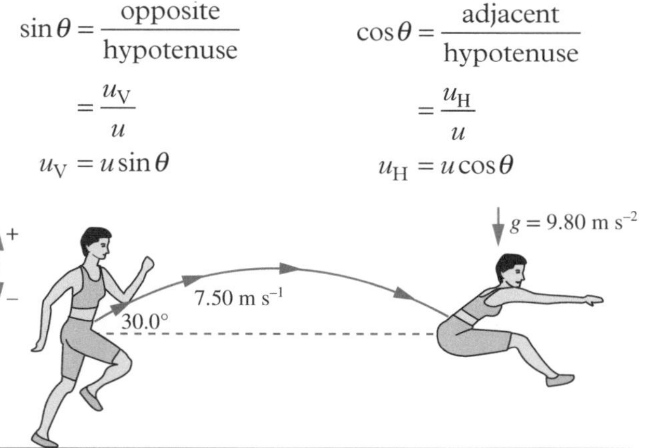

FIGURE 5.3 Projectiles can also be launched obliquely (at an angle).

As with projectiles launched horizontally, if air resistance is negligible, the only force that is acting on a projectile is gravity, F_g. Therefore, the horizontal velocity of a projectile remains constant throughout its flight, i.e. $v_{av} = \frac{s}{t}$ is used. At its highest point, the projectile is only moving horizontally so the vertical component of the velocity is zero at this point. Overall, the assumptions that must be made when analysing a projectile problem are as follows.

- Air resistance is negligible.
- If an object is dropped, $u = 0\,\text{m}\,\text{s}^{-1}$.
- At the highest point of a projectile's path, $v_V = 0\,\text{m}\,\text{s}^{-1}$.
- Final velocity = negative initial velocity, i.e. $u = -v$.

In the example given in Figure 5.3, in order to find the maximum height reached, you would first need to break the initial velocity into its vertical and horizontal components.

$$u_V = u \times \sin\theta$$

$$u_V = 7.50 \times \sin 30 = 3.75\,\text{m}\,\text{s}^{-1}$$

The vertical displacement can then be calculated using the formula $v^2 = u^2 + 2as$:

$$0^2 = 3.75^2 + 2 \times -9.8 \times s$$

$$s = 0.717\,\text{m upwards}$$

This means that the centre of mass of the person jumping will travel upwards to a maximum of 0.7 m above their starting height. If the centre of mass is originally 1 m off the ground, then the maximum height reached will be 1.7 m.

Circular motion

Circular motion involves the movement of an object along a circular path. It can be uniform (with a constant rate of rotation and speed), or non-uniform (with a changing rate of rotation).

CIRCULAR MOTION IN A HORIZONTAL PLANE

Uniform circular motion is motion along a circular path in which there is no change in speed, only a change in direction. For example, consider an athlete in a hammer throw event, swinging a steel ball in a horizontal circle with a constant speed (Figure 5.4). The velocity of the hammer at any instant is tangential (at a tangent) to its path. At one instant (point A), the hammer is travelling at $25\,\text{m}\,\text{s}^{-1}$ north, then an instant later (point B) at $25\,\text{m}\,\text{s}^{-1}$ west, then $25\,\text{m}\,\text{s}^{-1}$ south (point C), and so on.

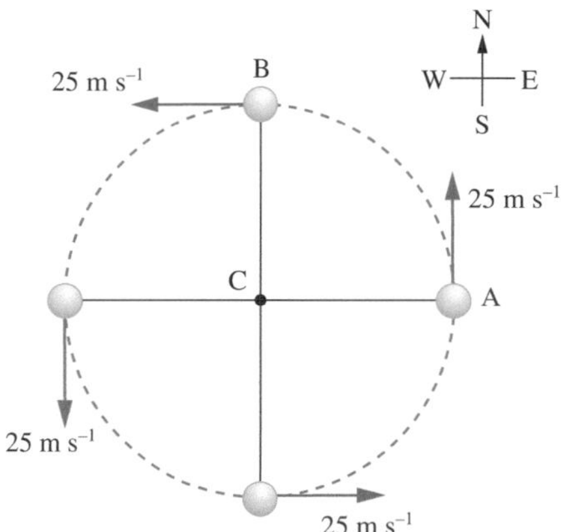

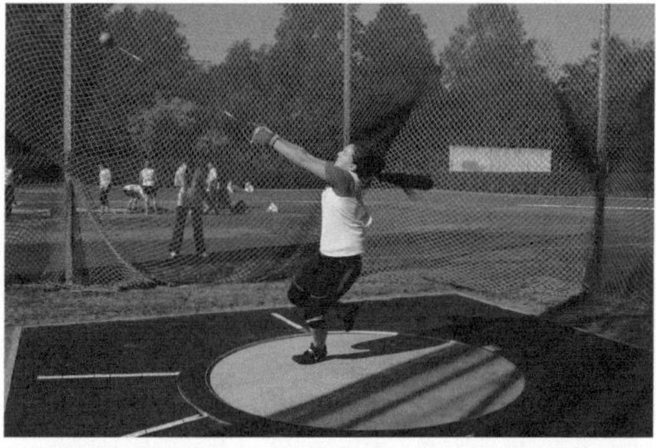

FIGURE 5.4 The velocity of the hammer (steel ball) at any instant is tangential to its path and is continually changing even though it has constant speed. This changing velocity means that the hammer is accelerating.

Period, T, is the time for one revolution and is measured in seconds. **Frequency**, f, is the number of revolutions each second and is measured in hertz (Hz). The relationship between T and f is:

$$f = \frac{1}{T} \text{ and } T = \frac{1}{f}$$

An object moving with a uniform speed in a circular path of radius, r, and with a period, T, has an average speed that is given by:

$$v = \frac{2\pi r}{T}$$

Sometimes it is more convenient to measure the angle of rotation, denoted by the angular velocity, ω. The **angular velocity** decribes the angle an object travels through, $\Delta\theta$, over a time t:

$$\omega = \frac{\Delta\theta}{t}$$

The angle can be found in either degrees or radians so that the angular velocity has units of either $^\circ\ s^{-1}$ or rad s^{-1}.

In a unit circle, the circumference is 2π units. Because of this, angles (in degrees) can be related to the arc length (in radians) on a unit circle. In a complete circle, the angle 360° is equal to 2π radians.

Therefore degrees can be converted to radians by multiplying by $\frac{\pi}{180}$, and radians can be converted to degrees by multiplying by $\frac{180}{\pi}$.

Because an object moving in a circular path is always changing its direction of motion, even if it is travelling with a constant speed it must have a non-zero acceleration. This acceleration is directed towards the centre of the circular path and is called **centripetal acceleration**, a_c:

$$a_c = \frac{v^2}{r}$$

Centripetal acceleration is a consequence of a **centripetal force** acting to make an object move in a circular path. Centripetal forces are directed towards the centre of the circle and their magnitude can be calculated by using Newton's second law for the net force of the object:

$$F_c = \frac{mv^2}{r}$$

By substituting the relationship between speed and period into the equations for acceleration and the net force, you can also find the following: $a_c = \frac{4\pi^2 r}{T^2}$ and $F_c = \frac{4\pi^2 rm}{T^2}$.

Centripetal force is always supplied by a real force, the nature of which depends on the situation. The real force is commonly friction, gravitation or the tension in a string or cable.

CIRCULAR MOTION ON BANKED TRACKS

This section focuses on circular motion on a surface that is not horizontal. On many road bends, the road is not horizontal, but is at a small angle to the horizontal. This is called a banked track, and it enables vehicles to travel at higher speeds when cornering, without skidding, than they could travel on a horizontal curved path. Banked tracks reduce a vehicle's dependence on friction to safely navigate a curve. An example of this effect can be seen at a cycling velodrome like that shown in Figure 5.5, where the banked corners allow cyclists to travel at much higher speeds than if the track were flat.

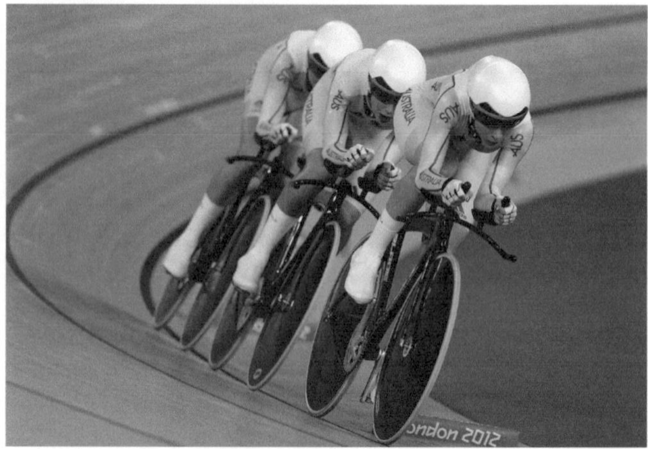

FIGURE 5.5 The Australian women's pursuit track cycling team in action on a banked velodrome track during the London Olympics in 2012

Banking a track eliminates the need for a sideways frictional force to turn. When cars travel in circular paths on horizontal roads, they are relying on the force of friction between the tyres and the road to provide the sideways force that keeps the car turning on a circular path. Both the normal force, F_N, and weight of the car are balanced, so the only force allowing the car to turn is friction. A banked track means that the normal force has an inwards component, enabling the car to turn the corner (Figure 5.6).

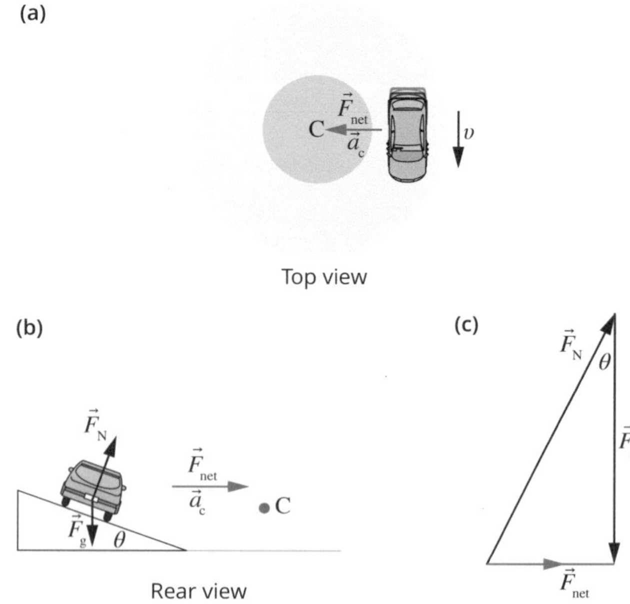

FIGURE 5.6 (a) The car is travelling in a circular path on a banked track. (b) The acceleration and net force are towards the centre of the circular motion, C. The banked track means that the normal force ($\vec{F}_N$) has an inwards component. This is what enables the car to turn the corner. (c) Vector addition gives the net force ($\vec{F}_{net}$) as acting horizontally towards the centre.

When the speed and angle are such that there is no sideways frictional force, the speed is known as the **design speed**. The forces acting on a vehicle travelling

ISBN 978 1 4886 1936 6

at the design speed on a banked track are gravity and the normal force from the track. These forces are unbalanced and add to give a net force directed towards the centre of the circular motion.

At the design speed, the angle of bank of the track can be determined by the following formula:

$$\theta = \tan^{-1}\left(\frac{v^2}{rg}\right)$$

This can therefore be rearranged to find the design speed:

$$v = \sqrt{rg\tan\theta}$$

A similar example of circular motion is the conical pendulum. A conical pendulum is a mass that travels in a horizontal circle on a string. The tension in the string supplies the force to counter the force of gravity and also supplies the centripetal force. Its construction is like an ordinary pendulum; however, instead of swinging back and forth, the mass of the conical pendulum moves at a constant speed in a circle with the string tracing out a cone. The closer the mass is to the horizontal, the more the speed increases. An example of a conical pendulum is shown in Figure 5.7.

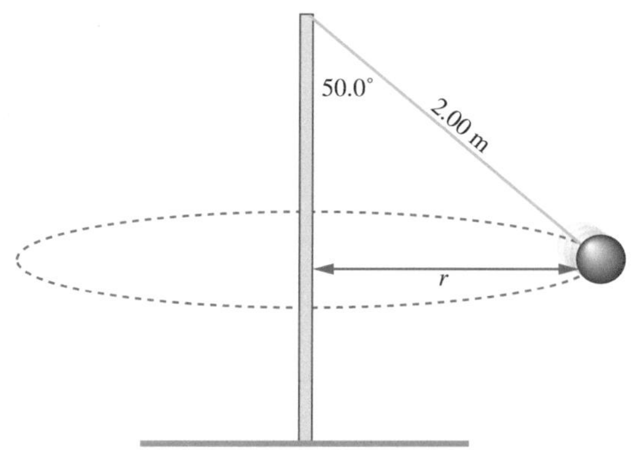

FIGURE 5.7 Conical pendulum

If both the angle and length of the string are known, the radius, r, of the circle can be found using a simple trigonometric ratio (see Figure 5.8a). Trigonometric ratios can also be used to find the other forces involved if one of the forces is known.

For example, if the mass of the ball is 0.20 kg, the weight force acting on it will be

$$F_g = mg = 0.20 \times 9.80 = 1.96\,\text{N downwards.}$$

Then as the weight of the ball (considering the string to be negligible mass) and the angle the ball is swinging at are both known, trigonometry can then be used to determine the centripetal force and tension in the string (see Figure 5.8b).

WORK AND ENERGY

Vertical circular paths (non-uniform circular motion), like horizontal circular paths, have an acceleration that is directed towards the centre of the circle. Unlike horizontal paths, however, in vertical paths the speed, as

(a)

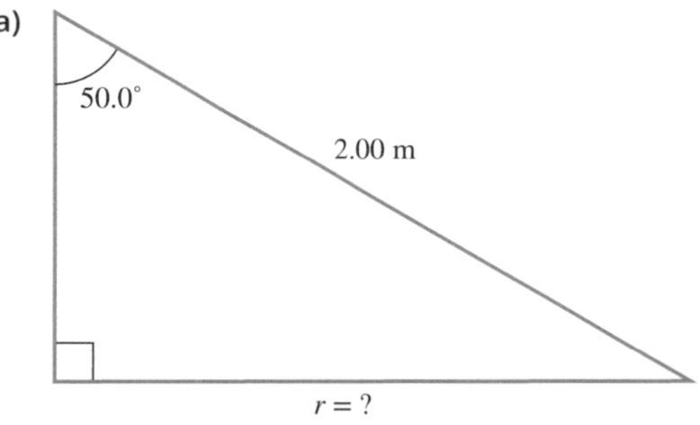

(b)

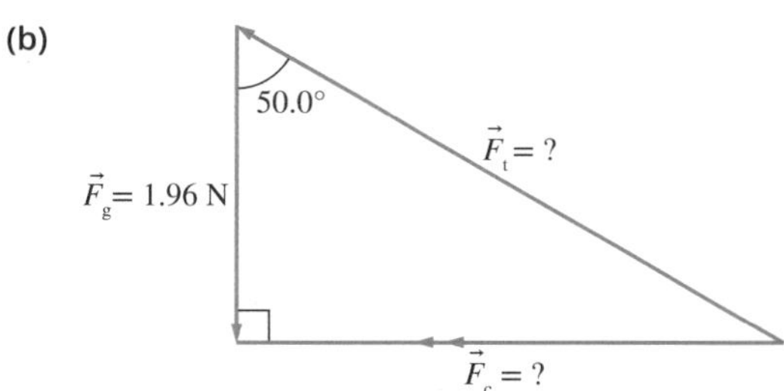

FIGURE 5.8 Trigonometry can be used to find (a) the radius of the circular motion as well as (b) the different forces acting: the weight force F_g, the tension of the string F_t and the net or centripetal force F_c.

well as the direction, of the object is constantly changing. For example, when travelling on a rollercoaster ride, strong forces would be experienced when travelling both through dips and over humps, but how these forces feel would be different. When you travel through dips, you feel as though you are getting pushed down into the seat, but when you travel over humps you feel as though you are being lifted up. This phenomenon is due to unbalanced forces between your weight and the normal force. Figure 5.9a shows the rider going through a dip. In this example, the centripetal force is upwards, which means the magnitude of the normal force is greater than the weight, and the rider will feel heavier than usual. Figure 5.9b shows the rider going over a hump. In this example, the centripetal force is downwards, so the magnitude of the normal force is less than the weight of the person and the rider will feel lighter than usual. The normal force is equal to the person's apparent weight, and this makes the person 'feel' heavier and lighter as they travel through the dips and humps respectively.

(a)

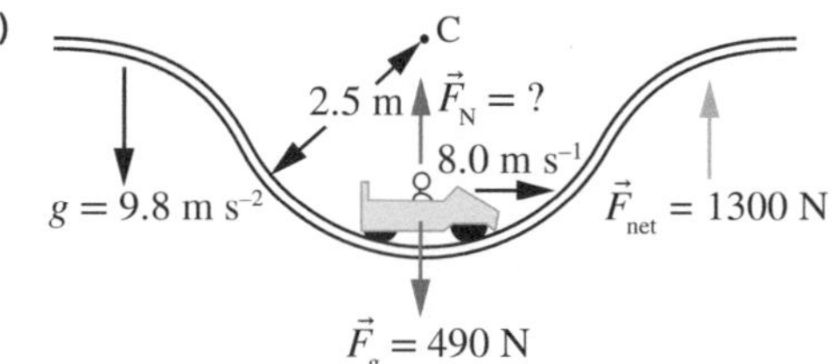

(b)

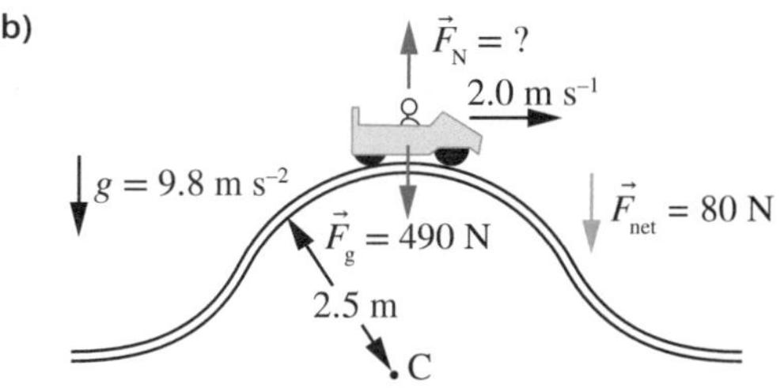

FIGURE 5.9 In (a), the centripetal acceleration is upwards towards the centre of the circle. In (b), the centripetal acceleration is downwards so the net force is also in that direction. At this point, the magnitude of the normal force $\vec{F}_N$ is less than the weight $\vec{F}_g$ of the person and the car.

As the speed is constantly changing throughout a roller coaster ride, often it is much easier to solve these types of problems using the law of **conservation of energy**. The law of conservation of energy states that energy is transferred or transformed between objects and there is always the same amount of energy at the end as there was at the start. The sum of the kinetic and potential energy (the total **mechanical energy**) of an isolated system is always conserved. This can be represented by the following:

$$(E_m)_{initial} = (E_m)_{final}$$

$$K_{initial} + U_{initial} = K_{final} + U_{final}$$

where:

E_m is the total mechanical energy

K is the kinetic energy

U is the potential energy.

Kinetic energy, K, is the energy of motion of a body:

$$K = \frac{1}{2}mv^2$$

where:

m is the mass (kg)

v is the speed (m s^{-1}).

Close to the surface of the Earth, where the force of gravity can be assumed to be constant, the change in gravitational potential energy of an object of mass m is dependent on the height from a given surface Δh and the acceleration due to gravity g:

$$\Delta U = mg\Delta h$$

The variable Δh is always given against some reference level, so gravitational potential energy can be positive or negative. A negative potential energy just means that the object will have travelled below the reference level. Ground level is often used as the reference point.

Consider the roller coaster ride again. There are times when the car isn't moving so that the kinetic energy is zero and you can describe the total energy of the system using only the gravitational potential energy. If the total energy in the system is known, this can be used to find the kinetic energy at other times during the ride as a change in potential energy is transformed into a change in kinetic energy, which means the speed can be found. Then the speed can be substituted into the centripetal force formula to calculate the net force.

Work is the transfer of energy from one object to another and/or the transformation of energy from one form to another (i.e. $W = \Delta U = \Delta K$). Work, W, is a scalar variable and is measured in joules (J). A force does work on an object when it acts on a body causing a displacement in the direction of the force. A centripetal force does no work on an orbiting object, as the force and displacement are perpendicular.

TORQUE

Torque is a measurement of the tendency of a force to cause an object to rotate around an axis. This axis is known as the **axis of rotation**. The object doesn't need to be circular for it to rotate around the axis of rotation. For example, torque is applied to a nut or bolt that needs to be tightened with a spanner. In this example, the pivot point is the bolt and the force is applied at right angles to the spanner (Figure 5.10). For all turning objects, there must be a **pivot point** around which the object will rotate, and there must be a force applied to the object to cause it to rotate. There also must be some distance between the **line of action of the force** (an imaginary line through the force vector) and the pivot point.

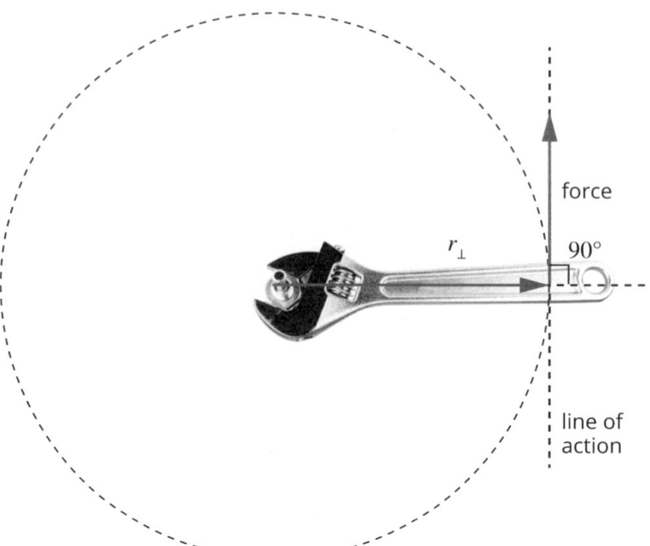

FIGURE 5.10 Although the adjustable spanner is not a wheel or circle, torque can still be applied to the nut.

The formula for calculating torque is:

$$\tau = r_\perp F$$

where:

τ is the torque (N m)

$r_\perp$ is the perpendicular force arm (m)

F is the force (N).

The amount of torque created is therefore directly proportional to the perpendicular distance between the pivot point and the line of action of the force, and magnitude of the force. This perpendicular distance is called the force arm. The force arm is given the symbol $r_\perp$ (as shown in Figure 5.10). Torque is a vector quantity. Generally the direction convention chosen describes a clockwise rotation as negative and an anticlockwise rotation as positive.

Maximum torque occurs when the acting force applied is perpendicular to a line drawn from the pivot point to the point of application. The larger the force acting on the object, the larger the torque will be. The longer the force arm, the greater the torque will be.

If torque is generated by an acting force that is not perpendicular to the lever arm of the object, then either:

- the component force perpendicular to the length of the object is used to calculate torque:

$$\tau = rF_\perp$$

or

- the distance from the pivot point perpendicular to the line of action of the force is used to calculate torque:

$$\tau = r_\perp F.$$

 ISBN 978 1 4886 1936 6

Both strategies for determining torque from non-perpendicular situations equate to:

$$\tau = rF \sin\theta$$

where θ is the angle between the applied force and the radius (°).

For example, consider someone loosening a bolt with a force of 65.0 N at an angle of 68.0° to the spanner, shown in Figure 5.11. In this example, the perpendicular force could first be found: $F_{\perp} = F \sin\theta = 65.0 \sin 68.0 = 60.3$ N. This can then be used to find the torque using $\tau = rF_{\perp}$. Alternatively, $\tau = rF \sin\theta$ can be used.

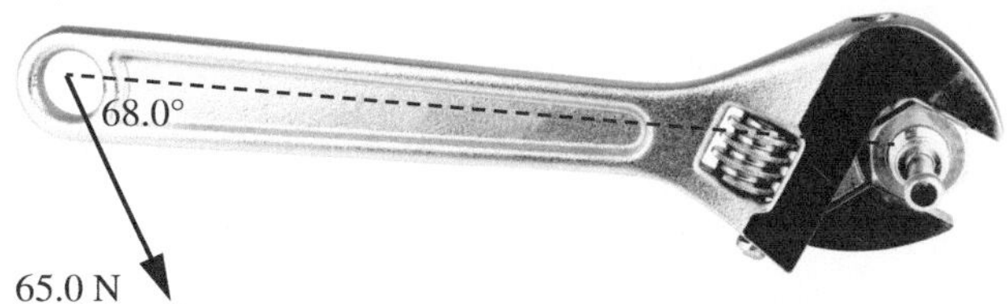

FIGURE 5.11 An acting force that is not perpendicular to the lever arm of the object will also generate a torque.

Motion in gravitational fields

Gravity is the weakest of the four fundamental forces of physics (electromagnetism, the strong nuclear force and the weak nuclear force are the others), but is still a very important force that drives the universe. For a satellite to move in an orbit around the Earth, the centripetal force is supplied by the gravitational force of attraction between the two masses. The size of this force is determined by one of Newton's laws: Newton's law of universal gravitation.

GRAVITY

All objects with mass attract one another with a gravitational force. The gravitational force acts equally on each of the masses. The magnitude of the gravitational force is given by **Newton's law of universal gravitation**:

$$F = \frac{GMm}{r^2}$$

where:

F is the gravitational force (N)

M is the mass of object 1 (kg)

m is the mass of object 2 (kg)

r is the distance between the centres of M and m (m)

G is the **gravitational constant**, $6.67 \times 10^{-11}\,\text{N}\,\text{m}^2\,\text{kg}^{-2}$.

The gravitational force is always an attractive force. Gravitational forces are usually negligible unless one of the objects is massive, e.g. a planet. For example, consider a man with a mass of 90 kg and a woman with a mass of 75 kg who have a distance of 80 cm between their centres. The force of gravitation would be:

$$F = \frac{GMm}{r^2}$$

$$= 6.67 \times 10^{-11} \times \frac{90 \times 75}{0.80^2}$$

$$= 7.0 \times 10^{-7}\text{ N towards one another.}$$

The weight of an object on the Earth's surface is due to the gravitational attraction of the Earth. A **gravitational field** is a region in which a gravitational force is exerted on all matter within that region. Some systems like the solar system involve several objects (i.e. the Sun and planets) exerting attractive forces on each other at the same time. Every physical object has an accompanying gravitational field. A gravitational field can be represented by a gravitational field diagram, in which:

- the arrowheads indicate the direction of the gravitational force
- the spacing of the lines indicates the relative strength of the field. The closer the line spacing, the stronger the field.

For example, Figure 5.12 shows a gravitational field diagram. The arrows indicate that objects will be attracted towards the mass in the centre and the spacing of the lines shows that force will be strongest at the surface of the central mass and weaker further away.

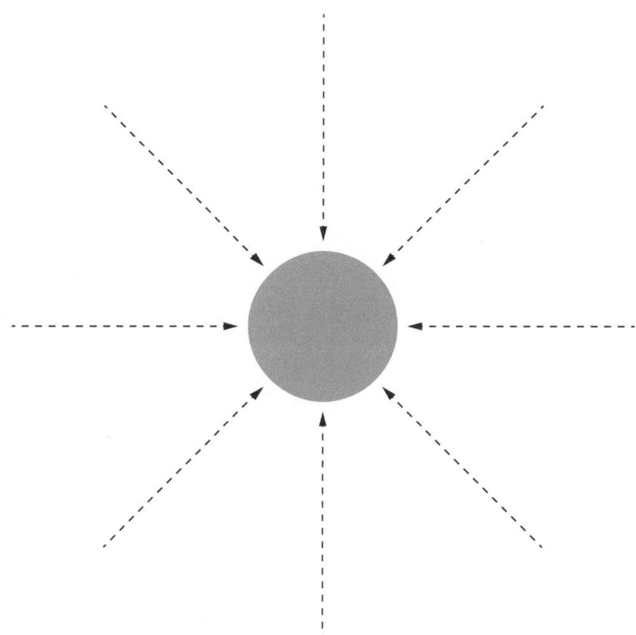

FIGURE 5.12 Gravitational field lines

The strength of a gravitational field can be calculated using the following formulae:

$$g = \frac{F_g}{m} \text{ or } g = \frac{GM}{r^2}$$

The acceleration due to gravity of an object near the Earth's surface can be calculated using the dimensions of the Earth:

$g = \frac{Gm_{Earth}}{(r_{Earth})^2} = 9.80\,\text{N}\,\text{kg}^{-1}$ towards the centre of the Earth.

This value for g varies from location to location and with altitude (but is consistent enough to be considered a constant for most Year 12 calculations for altitudes close to the Earth's surface). The gravitational field strength on the surface of any other planet depends on the mass and radius of the planet.

SATELLITE MOTION

A **satellite** is an object that is in a stable orbit around a larger central mass. Satellites can either be **natural** (e.g. the moon) or **artificial** (e.g. a communication satellite). Artificial satellites are very important for science, industry, communications and the military. The only force acting on a satellite is the gravitational attraction between it and the central body. Satellites are in continual free-fall. As satellites are moving at a velocity relative to the Earth, they are moving across the sky at the same rate that they are falling, so the combined effect is that they fall in a curved path. This effect is shown in Figure 5.13, which shows an astronaut orbiting the Earth.

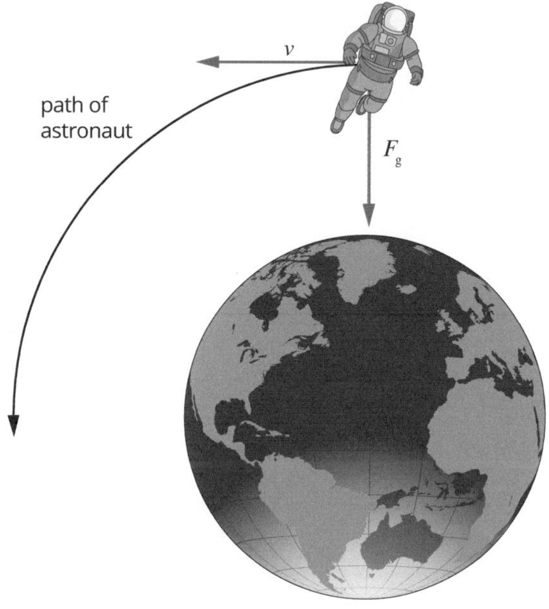

FIGURE 5.13 Astronauts are in free fall while orbiting the Earth.

Satellites move with a centripetal acceleration that is equal to the gravitational field strength at the location of their orbit. Artificial satellites are often equipped with tanks of propellant that are squirted in the appropriate direction when the orbit of the satellite needs to be adjusted. In a circular orbit, a satellite travels at a constant speed and stays the same distance from the Earth.

The speed of a satellite, v, in a circular orbit is given by:

$$v = \frac{2\pi r}{T}$$

This formula can then be used to determine the centripetal acceleration acting on this satellite. Using the scalar variables for these equations, the magnitude of the acceleration is then:

$$a_c = \frac{v^2}{r} = \frac{4\pi^2 r}{T} = \frac{GM}{r^2} = g$$

These relationships can be manipulated to determine any feature of a satellite's motion: its speed, radius of orbit or period of orbit. The magnitude of the gravitational force acting on a satellite in a circular orbit is given by:

$$F_c = \frac{mv^2}{r} = \frac{4\pi^2 rm}{T^2} = \frac{GMm}{r^2} = mg$$

Artificial satellites usually move in circular orbits; however, it does depend on the energy and trajectory. Satellites could follow a spiral, hyperbolic, elliptical or circular orbit. Johannes Kepler, a German astronomer, published three laws regarding the motion of planets. He was the first to work out that planets do not travel in circular paths, but rather in elliptical paths. His three laws are as follows.

- The planets move in elliptical orbits with the Sun at one focus.
- The line connecting a planet to the Sun sweeps out equal areas in equal intervals of time (Figure 5.14).
- For every planet, the ratio of the cube of the average orbital radius, r, to the square of the period, T, of revolution is the same, i.e. $\frac{r^3}{T^2}$ = a constant, k.

For any central body of mass, M:

$$\frac{r^3}{T^2} = \frac{GM}{4\pi^2}$$

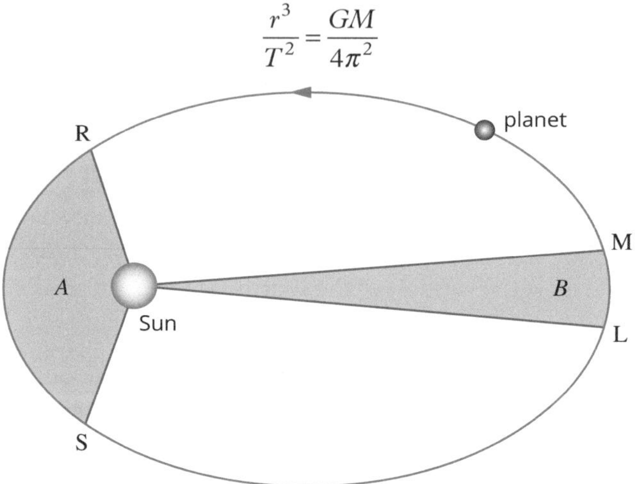

FIGURE 5.14 A line joining a planet to the Sun will sweep out equal areas in equal times. If the time taken for the planet to travel from R to S is the same as it takes to travel from M to L, then area A will equal area B.

If another satellite's orbital radius, r, is known this enables a satellite in the same system, with period T, to be determined.

GRAVITATIONAL POTENTIAL ENERGY

When considering gravitational potential energy with the motion of objects like rockets or satellites, a larger understanding is necessary. The gravitational potential energy formula, $\Delta U = mg\Delta h$, assumes that the Earth's gravitational field is constant. This is approximately true for objects that are within a few kilometres of the Earth's surface. Gravitational potential energy is important when combined with kinetic energy, $K = \frac{1}{2}mv^2$, and conservation of mechanical energy to calculate the speed of a falling object.

The strength of the Earth's gravitational field decreases as altitude increases. This means it is not sufficient to assume a constant value for the Earth's gravitational field, when considering objects like satellites or moons that orbit at high altitudes. Using Newton's universal law of gravitation and substituting this into the gravitational potential energy formula derives:

$$U = -\frac{GMm}{r}$$

ISBN 978 1 4886 1936 6

Gravitational potential energy is measured against a reference level and a negative value means that it has moved below this level. By convention, the gravitational potential energy of a satellite is zero when it has escaped the gravitational field of the Earth. In most cases, the satellite will be close to Earth and in Earth's gravitational field, making the potential energy a negative value.

The total energy E in a non-constant gravitational field can be found by combining the law of conservation of energy, Newton's law of gravitation, centripetal force and Newton's second law. The formula is:

$$E = U + K = -\frac{GMm}{2r}$$

When an object like a rocket has enough kinetic energy to escape the Earth's gravitational field, it is said to have reached **escape velocity**.

Since an object that has escaped a gravitational field has a potential energy of $U = 0$, its energy will be entirely kinetic. Due to the conservation of mechanical energy, this kinetic energy is equal in magnitude to the object's initial gravitational potential energy when $K = 0$. This can be used to find the escape velocity—when $K = U$:

$$v_{esc} = \sqrt{\frac{2GM}{r}}$$

WORKSHEET 5.1

Knowledge review—straight-line motion

1 The motion of a ball rolling down a plank from rest is measured by marking its position each second. From $t = 3\,s$ to $t = 4\,s$, the ball travels 56 cm. Calculate the magnitude of the acceleration of the ball.

2 From a platform 24.5 m above the ground, a stone is thrown straight up at $19.6\,m\,s^{-1}$.

a How long will it be in the air before hitting the ground?

b Calculate its velocity as it hits the ground.

3 An Alfa Romeo sports car travelling at $22\,m\,s^{-1}$ west overtakes a bus travelling at $9\,m\,s^{-1}$ in the same direction. Find:

a the velocity of the car relative to the bus

b the velocity of the bus relative to the car

c the direction the bus appears to move in the car driver's frame of reference.

ISBN 978 1 4886 1936 6

4 A 540 kg cable car is suspended from the middle of a light cable, with each half of the cable making an angle of 12.0° to the horizontal. Draw a force diagram of the situation and calculate the size of the tension in the cable.

5 Consider a 100 g teacup sitting on a kitchen bench.

a What is its weight?

b What is the size of the force of the benchtop on the cup? In which direction does it act?

c Explain why these forces do *not* make an action–reaction pair. What are the action–reaction pairs here?

6 An 8.0 kg box rests on a plank which is inclined at an angle of 20° to the horizontal. The coefficient of friction between the plank and the box, μ, is 0.40. If the plank is gradually raised, find the angle at which the box will start to slip down the plank. (Hint: You will need to find the size of the normal force first.)

7 The velocity–time graphs for three objects, A, B and C, dropped from the same height are shown below.

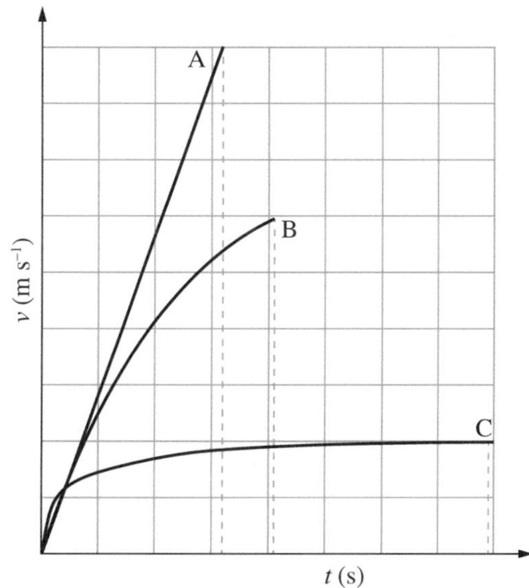

a What physical quantity is represented by the gradient of graph A?

b What do the areas underneath the graphs represent? How do they compare with each other?

c Explain what is happening to object C.

8 The foundations of large buildings are often built on supports called 'piles' which are driven down into the ground by a pile-driver. A 4.50×10^3 kg pile-driver mass falls through a height of 40.0 cm before colliding with the top of a pile. The mass drives the pile 30.0 cm into the ground when both the pile and the mass come to rest.

a Calculate the pile-driver's loss of potential energy.

b Calculate the average resistance force of the ground.

 ISBN 978 1 4886 1936 6

WORKSHEET 5.2

Projectile motion—is it safe?

A group of students intends to fire a small missile with a velocity of $40.0\,m\,s^{-1}$ at an angle of 60.0° from the horizontal from a point on the oval. They intend to aim at the windowless wall of a building 45 m away from the launch site, so that there is no danger of the missile hitting anyone. The building is 36 m high.

1 Construct a diagram of the situation, showing all known variables.

2 Prove that the projectile will travel over the height of the building and hence that this situation is unsafe.

3 As this situation is unsafe, make two possible recommendations in order to make it safe. Show calculations to prove that your proposal is safe, and note any assumptions.

RATING MY LEARNING	My understanding improved	Not confident ◄——► Very confident ○ ○ ○ ○ ○	I answered questions without help	Not confident ◄——► Very confident ○ ○ ○ ○ ○	I corrected my errors without help	Not confident ◄——► Very confident ○ ○ ○ ○ ○

WORKSHEET 5.3

Projectile motion—human cannonball data analysis

Human 'cannonballs' used to be a common feature of circuses and fairs. Risk assessments and ever-increasing insurance premiums brought an end to them. More than 30 people have died over the years since the first cannonball, a 14-year-old girl, was launched at the Royal London Aquarium in 1877.

These days, some stunt performers still maintain the tradition. The human cannonball is fired from a specially designed cannon, and lands on a horizontal net or inflated bag. Amazingly, the official world record stands at 59.05 m.

The data below was collected from the launch of a human cannonball of mass 80 kg blasted towards the east from a cannon and landing in a safety net. The height of the net was equal to the launch height. The angle of launch was 45° and the initial speed was $28\,m\,s^{-1}$. In your calculations, ignore air resistance unless otherwise instructed and use $g = 9.80\,m\,s^{-2}$.

Time after launch (s)	Speed ($m\,s^{-1}$)
0	28.0
1.0	22.4
2.0	20.0
3.0	22.4
4.0	28.0

Air resistance (N)	Distance travelled (m)
0	0
400	50
600	100
700	150
780	200
830	250
850	300
870	350
890	400
900	450
900	500
900	550
900	600

 ISBN 978 1 4886 1936 6

WORKSHEET 5.3

1 On the set of axes provided, produce a graph of vertical acceleration versus time for the human cannonball for the 4.0 s worth of motion. Use up as positive and ensure that your graph is appropriately labelled.

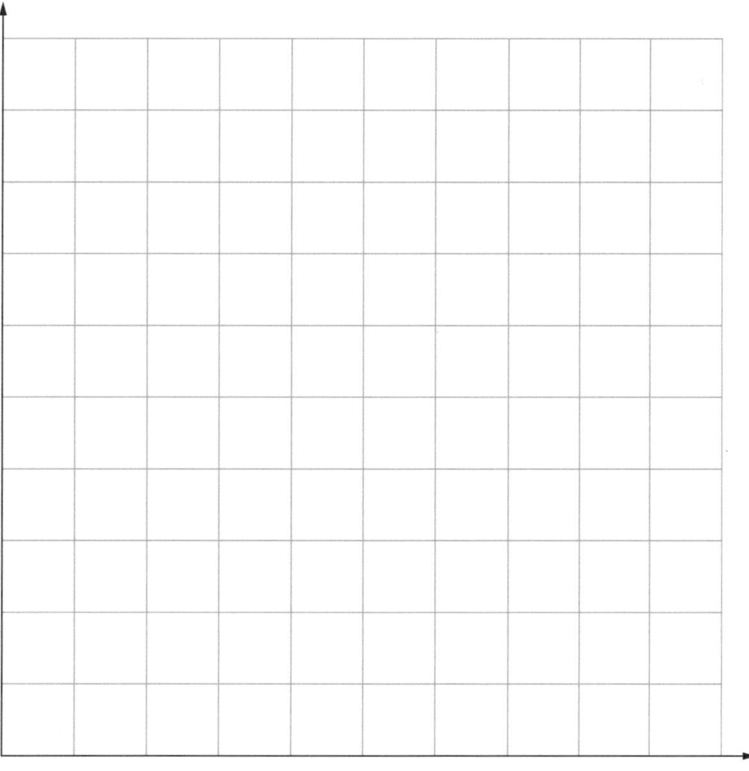

2 Using an energy approach, determine the:

a total mechanical energy of the human cannonball

b gain in gravitational potential energy of the human cannonball as they reach the maximum height

c change in height that the human cannonball achieved from launch.

3 Determine the height of the cannonball above the initial launch height after:

a 2.0 s

b 3.5 s.

4 Calculate the speed of the human cannonball after:

a 2.0 s

b 3.5 s.

5 What is the velocity and acceleration of the human cannonball at the maximum height?

6 Discuss the forces that are acting on the human cannonball when passing through the highest point in the trajectory if air resistance is ignored. State the magnitude and direction of the net force acting at this point.

7 Discuss the forces that are acting on the human cannonball when passing through the highest point in the trajectory if air resistance is taken into account.

RATING MY LEARNING	My understanding improved	Not confident ◄——► Very confident ○ ○ ○ ○ ○	I answered questions without help	Not confident ◄——► Very confident ○ ○ ○ ○ ○	I corrected my errors without help	Not confident ◄——► Very confident ○ ○ ○ ○ ○

 ISBN 978 1 4886 1936 6

WORKSHEET 5.4

Going around corners

In order for a body to travel in a circle or part of a circle, a centripetal force is required to be exerted on the object. This can be the pull of gravity for planets, the tension in a wire for a hammer thrower or the reaction force of a surface in contact with the object. In the case of a car going around a corner, the centripetal force must come from the sideways component of reaction from the road surface or the friction of the car tyres against the road surface.

The diagram below depicts a car travelling away from the viewer and turning to the left.

1 Draw and label the force vectors acting on the car at this instant.

2 Consider a car negotiating a suburban roundabout. The path taken has a radius of curvature of 8.0 m and the coefficient of friction between the tyre and the road is 0.72. What is the maximum speed the car can travel (in km h^{-1}) and still remain on the circular path?

3 The coefficient of friction between the tyre of a Formula One racing car and the asphalt track is typically 1.50. Calculate the minimum radius of curvature that can be taken by a Formula One car at 120 km h^{-1}.

4 Formula One cars are fitted with various aerofoils and body shaping, which can generate considerable downforce. Find how much downforce (as a fraction of the car's weight) would be needed to negotiate the same corner at a 30% greater speed.

5 When a motorcycle goes around a corner, the rider needs to lean over into the corner. On the diagram of a cornering motorcycle below, draw and label the forces acting on the motorcycle and the rider. (Point C, 0.65 m from the contact point, is the centre of mass of the bike and rider.)

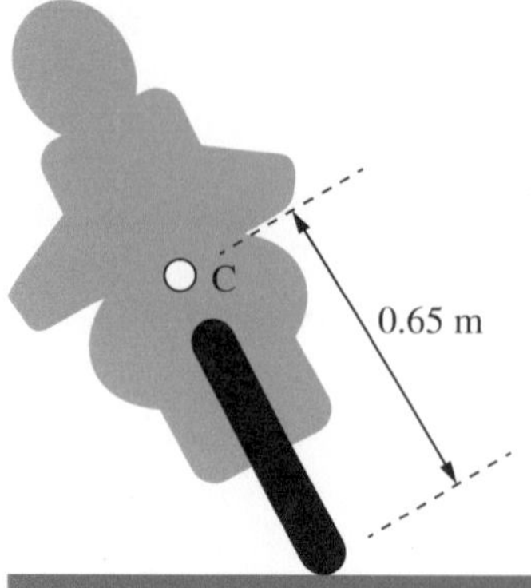

6 Assuming that the motorcycle's tyres can generate as much friction with the road as the car's tyres, find the angle of lean of the motorcycle from the vertical as it takes the corner.

Bicycles can travel faster around a bend by 'banking' the track as seen in a velodrome. The ideal situation on a banked curve is for the bike to be perpendicular to the track surface. The Dunc Gray Velodrome in Sydney was the venue for the track cycling in the 2000 Olympic Games. The maximum angle of bank in the end curves is 42°, while the straights are angled at 12.5°.

7 Anna Meares takes a path with a radius of curvature of 22 m where the track is banked at 42°. Find the greatest speed at which she can travel safely.

8 Find her angular velocity at that speed.

There are many further points to consider when investigating how vehicles can be helped to negotiate corners.

- When a car first enters a bend and when it leaves the bend, it is usually travelling faster than in the middle of the bend. This means the radius of curvature of the road needs to vary, and should be smallest at the apex of the corner.
- The radius of curvature of a velodrome track also changes as the banking angle varies from the straights to the ends.
- The greater the angle of bank, the harder a car is 'pushed' down onto the track. If pushed too far, the car body will touch the track and the tyres may lose adhesion.
- If a cyclist leans on a banked curve, they can generate more friction and take the curve even faster—provided they can pedal fast enough!

RATING MY LEARNING	My understanding improved	Not confident ◄—► Very confident ○ ○ ○ ○ ○	I answered questions without help	Not confident ◄—► Very confident ○ ○ ○ ○ ○	I corrected my errors without help	Not confident ◄—► Very confident ○ ○ ○ ○ ○

 ISBN 978 1 4886 1936 6

WORKSHEET 5.5

Circular motion and gravity

When a rocket takes off, it accelerates quickly to achieve enough speed to enter orbit. In a related situation, fighter pilots will undergo significant '*g*-forces' while performing tight turns. To prepare themselves for this, the astronauts and pilots will need to experience high '*g*-forces' during their training.

1 A brief period of acceleration can be achieved using a rocket sled equipped with a jet turbine or rocket motor. Consider a person strapped into such a sled where the total mass is 450 kg. The rocket motor exerts an upwards force of 36.0 kN for a 4.00 s burn. Calculate the acceleration produced and the distance travelled during the burn.

2 The acceleration can be measured in units of g, where $g = 9.80\,\text{m s}^{-2}$ at or near the Earth's surface. How many g's does the passenger experience?

Another way of producing high-g conditions is to use a centrifuge—a seat or capsule attached onto the end of a long arm which can be rotated at a variable rate in a horizontal plane.

3 A centrifuge with an 8.5 m arm is rotated at 24 rpm. How many g's are experienced by the test pilot in the cage at the end of the arm?

Anybody who is in orbit experiences 'apparent' weightlessness. Of course, they are not really weightless at all as they are still in the Earth's gravitational field—the very reason they are in orbit. However, they feel weightless because there are no contact forces from the sides of the spacecraft acting on them. As part of their training before they go to space, astronauts need to get experience of weightlessness. This is where the 'Vomit Comet' comes in.

There are several aeroplanes which qualify for this dubious title. Each plane follows a series of parabolic, free-fall segments during its flight. These free-fall sections are joined together by a series of dives and climbs during which higher than normal '*g*-forces' are experienced by the passengers. A typical flight profile is as shown here.

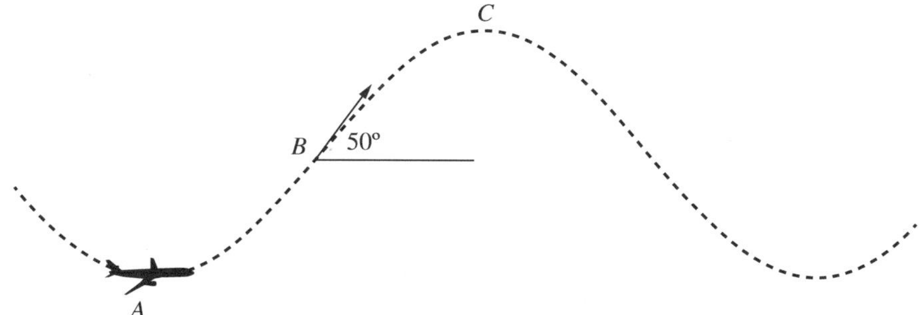

ISBN 978 1 4886 1936 6

4 Consider point A at the bottom of the dive phase where the aeroplane is travelling at $720\,km\,h^{-1}$. A 65.0 kg person lying on the floor of the cabin experiences 1.80*g*. Draw a diagram showing the forces acting on her.

5 Calculate the radius of curvature of the flight path at that point.

6 From that point the aeroplane begins to climb at an increasing angle until it reaches an elevation of 50° at point B. At this point it begins the parabolic trajectory phase by reducing engine power. Assuming the passenger still experiences 1.80*g* just before the engines are throttled back, draw a new diagram showing the forces acting on her then.

7 During this part of the flight, the engine power is reduced—but not to zero. If the aeroplane is entering free-fall, why does the plane still need the thrust of the engines?

8 Typically, the parabolic trajectory phase lasts 20 s: 10 s up and 10 s down. Calculate the velocity the plane will need at point B to achieve this duration of weightlessness.

9 Explain what would happen if the aeroplane were to fly a bit too fast at the top of the parabolic section of the flight at point C.

RATING MY LEARNING	My understanding improved	Not confident ◄——► Very confident ○ ○ ○ ○ ○	I answered questions without help	Not confident ◄——► Very confident ○ ○ ○ ○ ○	I corrected my errors without help	Not confident ◄——► Very confident ○ ○ ○ ○ ○

 ISBN 978 1 4886 1936 6

WORKSHEET 5.6

Working with orbits

Getting a satellite into space is not a trivial exercise. The objective is to achieve enough speed and height to maintain orbit. Consider the task of putting a 100 kg satellite into a circular orbit.

1 Given the radius ($r_E = 6.37 \times 10^6$ m) and mass ($m_E = 5.97 \times 10^{24}$ kg) of the Earth, calculate the gravitational potential energy of the satellite before it is launched.

2 Find the total energy of the satellite once it has been placed in an orbit of altitude 100 km above the Earth's surface.

3 Explain why these quantities are negative.

4 The difference between the two quantities you have found is the amount of energy that needs to be imparted to the satellite by the launch vehicle. How much energy is this?

5 Of course, the satellite before launch is not really stationary—it is moving along with the Earth as it rotates. Assuming that the launch pad is on the equator, find the initial kinetic energy of the satellite.

6 This initial kinetic energy can be utilised by launching the satellite in the appropriate direction. What is that direction and why is the equator the best latitude for launch?

7 Describe this kinetic energy as a percentage of the amount of energy found in question 4.

Now consider what is required to get such a satellite to Mars. To simplify things, first consider what energy is required to move the 100 kg satellite from Earth's orbit to the orbit of Mars.

8 Find the difference between the total energy of the satellite when it is in Earth's orbit compared to when it is in Mars' orbit. (Use this data: mass of the Sun $M_{\odot} = 1.99 \times 10^{30}$ kg, radius of Earth's orbit = 1.50×10^{11} m, radius of Mars' orbit = 2.28×10^{11} m.)

This energy difference is only a part of the story. In addition, the satellite will need to be given enough energy to escape entirely from Earth's gravity. (This is also known as escaping Earth's gravity well.) When the satellite is in Mars' orbit, its gravitational potential energy in Earth's field can be taken to be zero as it is so far away.

9 What is the value of this escape energy? What percentage of the energy calculated in question 8 does this represent?

Now consider the trajectory required for this orbital transfer. Rather than trying to travel to Mars in a straight line, the path will be determined by the relative positions of Earth and Mars, both at the time of launch and at the planned rendezvous with Mars.

The orbit required for greatest efficiency is called the Hohmann transfer orbit. Where the orbits of Earth and Mars are approximately circular, the transfer orbit is elliptical, with the perihelion (the closest point to the Sun) equal to the Earth's orbital radius and the aphelion (the furthest point from the Sun) at the orbit of Mars.

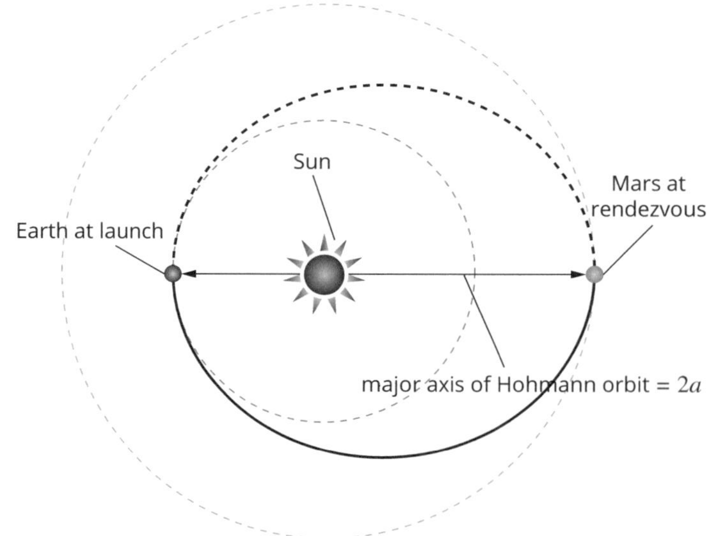

Kepler's third law can be used to find out roughly how long the trip to Mars would take. The third law states that for all orbits around the Sun the quantity $\frac{R^3}{T^2}$ is a constant, where R is the radius of the orbit and T is the period. Most planets have a circular orbit and the application of the third law involves the orbital radius. But when the orbit is highly elliptical, the value of R needs to be replaced with the length of the semi-major axis, a. The third law then becomes $\frac{a^3}{T^2}$ = constant.

 ISBN 978 1 4886 1936 6

10 Find the value of the semi-major axis of the Hohmann transfer orbit.

11 Compare the value of $\frac{a^3}{T^2}$ for the satellite in the Hohmann transfer orbit with the value of $\frac{a^3}{T^2}$ for the Earth, and find the duration of the Hohmann trip to Mars in units of days.

This exercise has presented some of the preliminary analyses required to plan a trip to Mars. There are many further questions that can be considered, including:

- Exactly when are both the Earth and Mars in the correct positions for a launch? (This is known as the launch window.)
- You have found the energy for the transfer orbit, but at what points does this energy need to be applied?
- How much does the orbital energy of the satellite in its 100 km orbit help?
- How will the gravity well of Mars affect calculations?
- Will a trip back to Earth be needed at some stage? (This has not yet been achieved.)

RATING MY LEARNING	My understanding improved	Not confident ◄—► Very confident ○ ○ ○ ○ ○	I answered questions without help	Not confident ◄—► Very confident ○ ○ ○ ○ ○	I corrected my errors without help	Not confident ◄—► Very confident ○ ○ ○ ○ ○

WORKSHEET 5.7

Calculating torque

Torque is the rotational equivalent of force. Torque must be applied to an object to start an object rotating. For calculations involving torque, the entire weight of an object can be considered to act at a single point, the centre of gravity or centre of mass of the object. Torque is a vector quantity with direction parallel to the axis of rotation.
A torque is positive when it is applied in the anticlockwise direction and negative when it is applied in the clockwise direction.

In an experiment to investigate torque and the rotation of mechanical systems, the following system was set up.

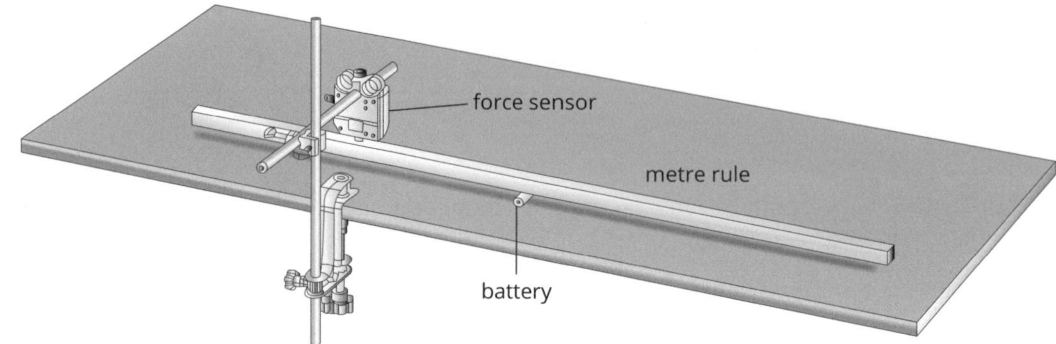

The battery acts as the pivot point about which the metre rule will rotate. The force sensor measures the force at a fixed point when a mass is placed at varying distances along the rule on the opposite side of the pivot point.

The following data was collected in a series of trials when the horizontal distance from the force sensor to the pivot point was 0.350 m.

Trial	Horizontal distance of mass to pivot point (m)	Force measured by sensor (N)	Torque measured by sensor (N m)	Torque applied by mass (N m)
1	0.100	0.280		
2	0.200	0.560		
3	0.300	0.840		
4	0.400	1.12		
5	0.500	1.40		

1 Assuming the rule is in rotational equilibrium, complete the table to show the torque measured by the sensor and the torque applied by the mass. Treat the direction of the torque measured by the force sensor as anticlockwise (i.e. imagine you are looking at the apparatus from across the table).

2 Based on the test results, what is the size of the mass being applied?

3 Draw a force diagram of the setup above when the net torque acting on the object is zero. Indicate the direction and magnitude of the forces involved.

 ISBN 978 1 4886 1936 6

4 The equipment is rearranged so that the force sensor is now at an angle of 60° from the horizontal. What would be the force measured by the force sensor with the mass at a distance of 0.500 m from the pivot point? Assume that the force sensor is still applied at a distance of 0.350 m from the pivot point.

5 Two children are sitting on a seesaw. If the child on the left has a mass of 31.0 kg and sits 3.0 m from the pivot point, and the child on the right has a mass of 25.0 kg and sits 4.0 m from the pivot point, what is the net torque on the seesaw? If the seesaw is initially at rest, which way will it rotate? Assume that the pivot point is directly below the centre of mass of the seesaw. Show your working.

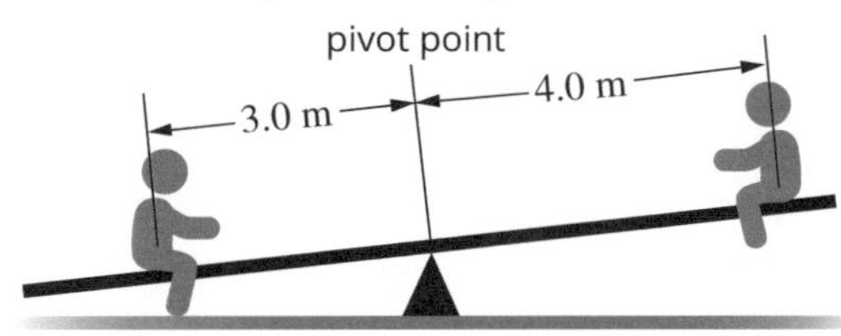

6 A log (m = 52 kg) has been placed on two supporting concrete blocks as shown. Assuming the log is in static equilibrium, find the normal force acting on each of the concrete blocks.

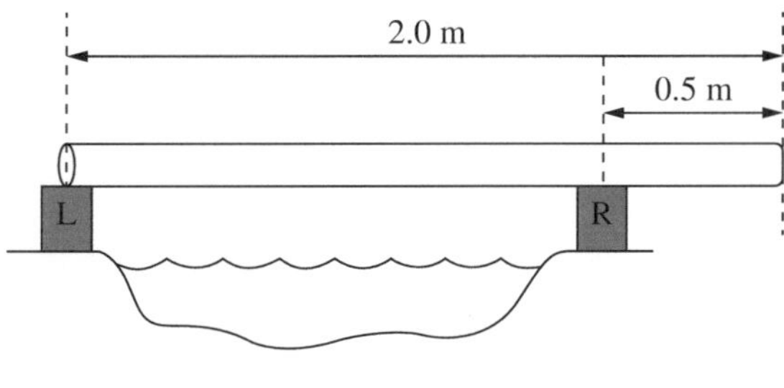

RATING MY LEARNING	My understanding improved	Not confident ◄──► Very confident ○ ○ ○ ○ ○	I answered questions without help	Not confident ◄──► Very confident ○ ○ ○ ○ ○	I corrected my errors without help	Not confident ◄──► Very confident ○ ○ ○ ○ ○

WORKSHEET 5.8

Literacy review—talking motion

Complete the statements by filling in the blanks from the list of randomly ordered words below. Some words may not be used and others may be used more than once.

gravity	free fall	speed	range	angle	circular	centripetal	parabolic
velocity	mass	projectile	radius	acceleration	constant	centrifugal	tension

1 When a ball falls vertically to the floor, it is in a ____________ state of motion. When a ____________ is launched horizontally, its horizontal velocity is ____________. The shape of the ball's flight is referred to as its trajectory and is ____________ in shape. The maximum ____________ of flight depends on the launch ____________ and the launch ____________. The force due to ____________ directly relates to the free fall acceleration of the ball.

2 The force that acts through a string is referred to as ____________. If an object is tied to a string such that its speed remains constant as it travels around a central point, it undergoes uniform ____________ motion. Even though the object's ____________ is constant, its ____________ is constantly changing. The tension in the string supplies the ____________ force.

3 An inwards, or centre-seeking, force is called the ____________ force. A constant force means a constant ____________, which makes sense if the object undergoing circular motion has a constantly changing ____________. If the centripetal force is kept constant then variables such as ____________ and ____________ are proportional to the time it takes to complete a revolution.

4 Centripetal and centrifugal forces are real and imaginary forces respectively that are used in discussions of circular motion. Find the derivation of each word and, from this, its definition.

____

____

5 Match each of the following terms with the description that best matches it.

Term	Definition
torque	
rotation	
lever arm	

a produced by a torque

b produces rotational motion

c distance from a turning point to the fulcrum

d direction of the applied force on a lever

RATING MY LEARNING	My understanding improved	Not confident ◄——► Very confident ○ ○ ○ ○ ○	I answered questions without help	Not confident ◄——► Very confident ○ ○ ○ ○ ○	I corrected my errors without help	Not confident ◄——► Very confident ○ ○ ○ ○ ○

 ISBN 978 1 4886 1936 6

WORKSHEET 5.9

Thinking about my learning

On completion of Module 5: Advanced mechanics, you should be able to describe, explain and apply the relevant scientific ideas. You should also be able interpret, analyse and evaluate data.

1 The table lists the key knowledge covered in this module. Read each and reflect on how well you understand each concept. Rate your learning by shading the circle that corresponds to your level of understanding for each concept. It may be helpful to use colour; for example:

- green—very confident
- orange—in the middle
- red—starting to develop.

Concept focus	**Rate my learning**				
	Starting to develop ◄——► Very confident				
equations of motion	○	○	○	○	○
projectile motion	○	○	○	○	○
circular motion in a horizontal plane	○	○	○	○	○
circular motion in a vertical plane	○	○	○	○	○
motion on banked tracks	○	○	○	○	○
torque	○	○	○	○	○
Newton's law of universal gravitation	○	○	○	○	○
satellite motion and Kepler's laws	○	○	○	○	○
escape velocity	○	○	○	○	○
energy changes of an object in circular motion and for satellites	○	○	○	○	○

2 Consider points you have shaded from starting to develop to middle-level understanding. List specific ideas that you found challenging.

3 Write down two different strategies that you will apply to help further your understanding of these ideas.

PRACTICAL ACTIVITY 5.1

Projectile motion—an introduction

Suggested duration: 40 minutes

INTRODUCTION

In this activity you will conduct a practical investigation to collect primary data to validate the relationship between time of flight, height and range.

PURPOSE

To confirm the dependence of the range of a projectile (the horizontal distance it travels) on its time of flight and launch velocity by predicting the landing point of a projectile and then testing the prediction.

MATERIALS

- ramp or board
- small ball (each group may use one or more of different size and mass)
- stopwatch
- horizontal bench or table
- polystyrene cup

PROCEDURE

Consider the situation in the figure below. A small ball rolls across a flat surface with a known speed, and then moves off the end of the table and falls to the floor. While it is in flight the only external forces acting on the ball are gravity and air resistance. Air resistance should be ignored in this investigation.

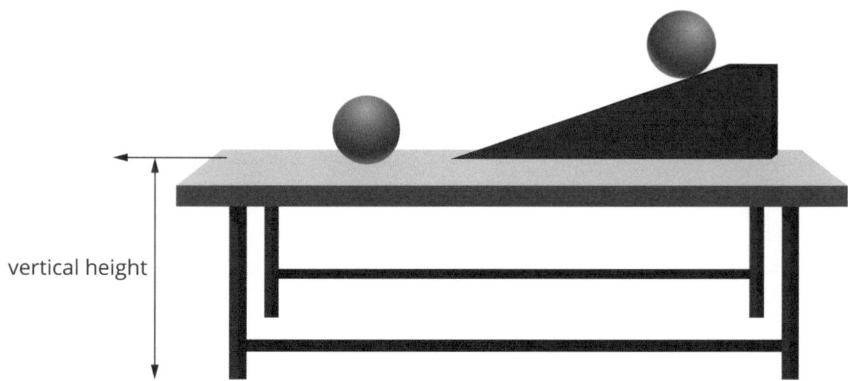

Part A: Determining the launch velocity

1 Mark a point on the ramp, where the ball can be consistently released from the same position. Measure the horizontal distance from the end of the ramp to the edge of the table. Record this in the Data and analysis section.

2 Release the ball and time how long it takes to travel the horizontal distance from the base of the ramp to the edge of the table. Do not let the ball fall to the floor while recording this time.

3 Repeat the test a number of times, recording each trial in Table 1 provided in the Data and analysis section.

4 Complete Table 1 by calculating the velocity across the tabletop for each trial and then finding the average velocity.

 ISBN 978 1 4886 1936 6

Part B: Predicting and testing the range

1 Record the height the ball will fall from the table to the floor.

2 Calculate the time of flight and hence the distance the ball will travel in the horizontal direction (the range) while it is in the air. Show your working in the Data and analysis section.

3 Measure out the calculated landing distance on the floor and place a polystyrene cup at the predicted point. Attach the cup to the floor with some Blu Tack so it won't move when the ball hits it or lands in it.

4 Test your prediction by releasing the ball from the spot marked on the ramp as originally used in Part A.

DATA AND ANALYSIS

Part A: Determining the launch velocity

Distance across tabletop: s = ______________ m

TABLE 1 Data collected for Part A

Trial number	Time, t, taken to travel across tabletop (s)	Velocity across tabletop ($m\,s^{-1}$) $v = \frac{s}{t}$
1		
2		
3		
4		
5		
average		

1 Is the velocity being calculated the velocity of the ball at the edge of the table? If not, is it a reasonable approximation? Explain your answer.

2 What effect would increasing the horizontal distance have on the reliability of your measurements?

Part B: Predicting and testing the range

Height above the ground: _______ m

1 Calculate the time it takes for the ball to fall from the table to the floor.

2 Calculate the distance the ball will travel in the horizontal direction, i.e. the range.

PRACTICAL ACTIVITY 5.1

CONCLUSION

1 State whether your prediction was successful, and describe any difficulties encountered in testing the prediction.

2 In this experiment, the assumption was made that there is negligible effect from air resistance. Would the effect of air resistance be more significant if the ball was released from a height of 30 cm up the ramp or 15 cm? Explain.

3 What is the major source of error in this experiment? What steps were taken to minimise it?

RATING MY LEARNING	My understanding improved	Not confident ◄──► Very confident ○ ○ ○ ○ ○	I answered questions without help	Not confident ◄──► Very confident ○ ○ ○ ○ ○	I corrected my errors without help	Not confident ◄──► Very confident ○ ○ ○ ○ ○

 ISBN 978 1 4886 1936 6

PRACTICAL ACTIVITY 5.2

Projectile motion—the effect of launch angle on range

Suggested duration: 45 minutes

INTRODUCTION

In this activity students will use electronic means to:

- measure the initial velocity of a projectile
- measure the time of flight of a projectile
- interpret data to predict the angle that will give the longest range
- calculate the appropriate angle in order to hit a target at a given range.

PURPOSE

To investigate the relationship between the launch angle of a projectile, its motion and the range of the projectile.

 **Safety warning: Always wear safety glasses when using any kind of projectile launcher. Never look down the barrel of a mechanical projectile launcher.**

MATERIALS

- data-collection system
- projectile launcher (commercial or improvised e.g. poly tube)
- projectile
- photogate/s and (optional) time-of-flight pad or stopwatch
- angle indicator
- tabletop or bench
- table clamp or burette stand and clamps
- A4 paper
- tape measure
- sticky tape
- carbon paper (optional)

PROCEDURE

1 Start a new experiment on your data collection system. Connect the photogates to your system following the manufacturer's instructions.

2 Select 'velocity between gates' if prompted by your data collection system.

Ensure the 'space between gates' parameter on your data collection system is set to the measured space between your photogates.

3 Put your data collection system into manual sampling mode with manually entered data. Name the manually entered numerical data 'angle' measured in degrees.

4 Add a column to the table to display the distance in metres.

5 Attach the projectile launcher to a table so that the projectiles travel across the longest part of the table. One suitable arrangement of the launcher is shown below. Use the equipment available to you to arrange the launcher to 'fire' down the length of the table and through the photogates. Be careful to avoid firing the projectile at classmates!

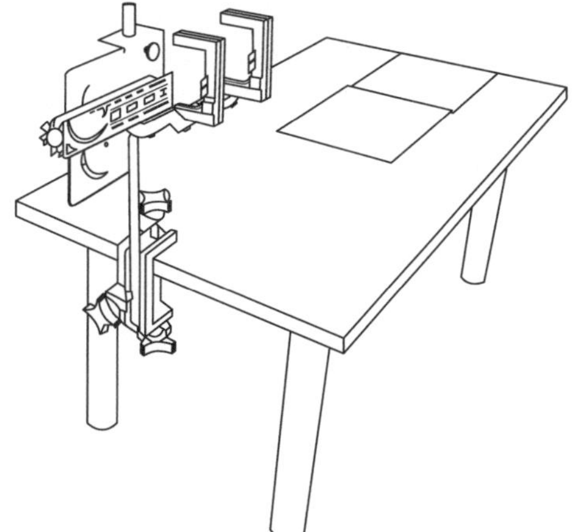

6 Place sheets of paper end-to-end in a line across the length of the table in front of the projectile launcher, and secure them in place with tape.

7 Measure the height from the point where the ball is released to the tabletop, and record this value in the Data and analysis section.

8 Mount the photogates to the launcher. Be sure to mount the first named photogate in your data analysis software closest to the launcher.

Part A: Distance versus angle

1 Set the launcher in the horizontal position with a launch angle of 0°.

2 Load a projectile into the launcher, and ensure that the launcher is set to its maximum compression or distance setting.

3 Launch the projectile, and note the point of impact on the paper.

4 Lay a sheet of carbon paper on top of the white paper over the point of impact, carbon side down, so that when a ball lands on it there will be a mark on the paper. Place a sheet of paper over the carbon paper to prevent damage to the carbon paper by the projectile.

   If carbon paper is not available, look for a small indentation on the paper where the ball hits. Highlight the point with a pencil or marker when the projectile lands. Before continuing, answer question 1 in the Data and analysis section.

5 Start recording with the data collection system and launch the projectile.

   Record the sampled 'velocity between gates' data point, and enter the corresponding angle value in Table 1 in the Data and analysis section.

6 Move the carbon paper, and measure the distance to the mark. Write the angle next to the mark on the paper.

7 Use the angle indicator on the launcher to position the launcher at the next angle, 10°.

8 Repeat the data collection steps, increasing the angle of inclination by 10° each time until you have recorded a data point every 10° from 0° to 80°.

9 Measure and enter the horizontal distance for each angle value into Table 1. Draw a graph of distance versus angle and a graph of velocity versus angle in the spaces provided.

Part B: Time of flight

1 If a time-of-flight pad is available, remove one photogate and attach the time-of-flight pad. Alternatively, a hand-held stopwatch or other timing mechanism can be used. Position the time-of-flight pad over the landing point recorded for an angle of 0° and reset the launcher to an angle of 0°. Before continuing, answer question 2 in the Data and analysis section.

2 Start a new data-collection session. Launch the ball from the launcher and, using the time-of-flight pad or a stopwatch, record the time the ball is in flight in Table 1.

3 Use the angle indicator on the launcher to position the launcher at the next angle, and repeat the data collection steps until you have recorded a data point every 10° from 0° to 80°. Record the time of flight for each angle in Table 1 of the Data and analysis section. Draw a graph of time of flight versus angle in the space provided.

DATA AND ANALYSIS

Ball release height: ________ m

1 What launch angle do you predict will yield the greatest range (horizontal distance)?

____

2 Which angle do you think will give the greatest time of flight? Explain the reasons behind your prediction.

____

____

 ISBN 978 1 4886 1936 6

TABLE 1 Collected data for your projectile

Angle (°)	Velocity (m s^{-1})	Distance (m)	Time of flight (s)
0			
10			
20			
30			
40			
50			
60			
70			
80			

Distance versus angle

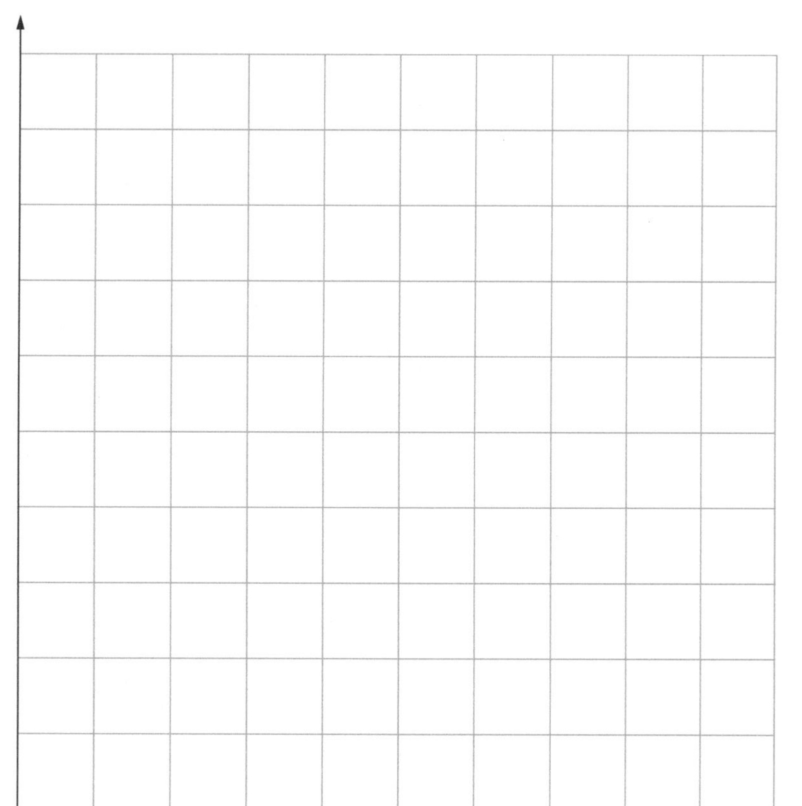

ISBN 978 1 4886 1936 6

PRACTICAL ACTIVITY 5.2

Initial velocity versus angle

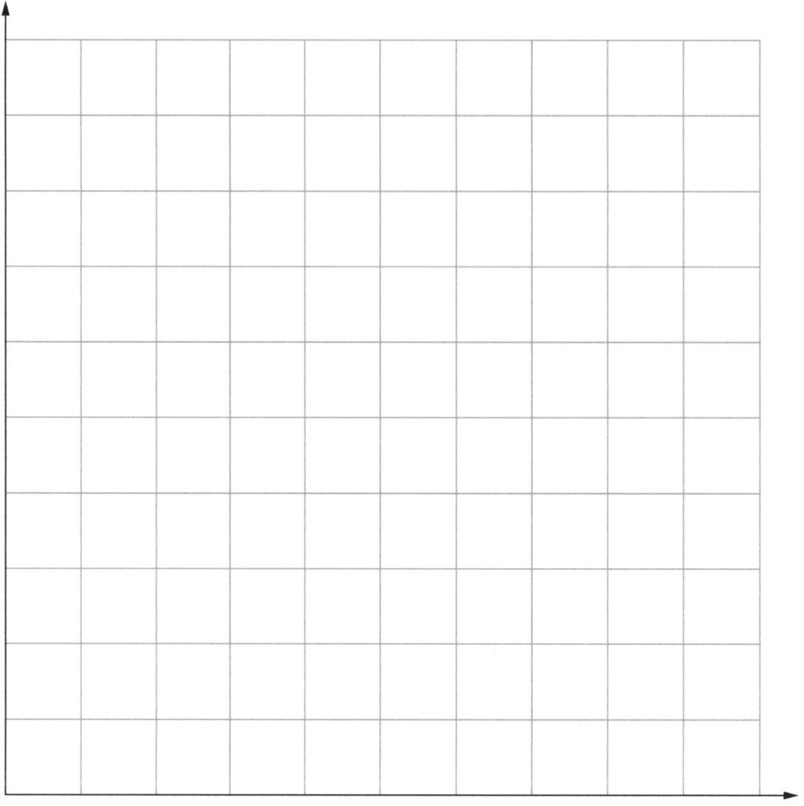

Time of flight versus angle

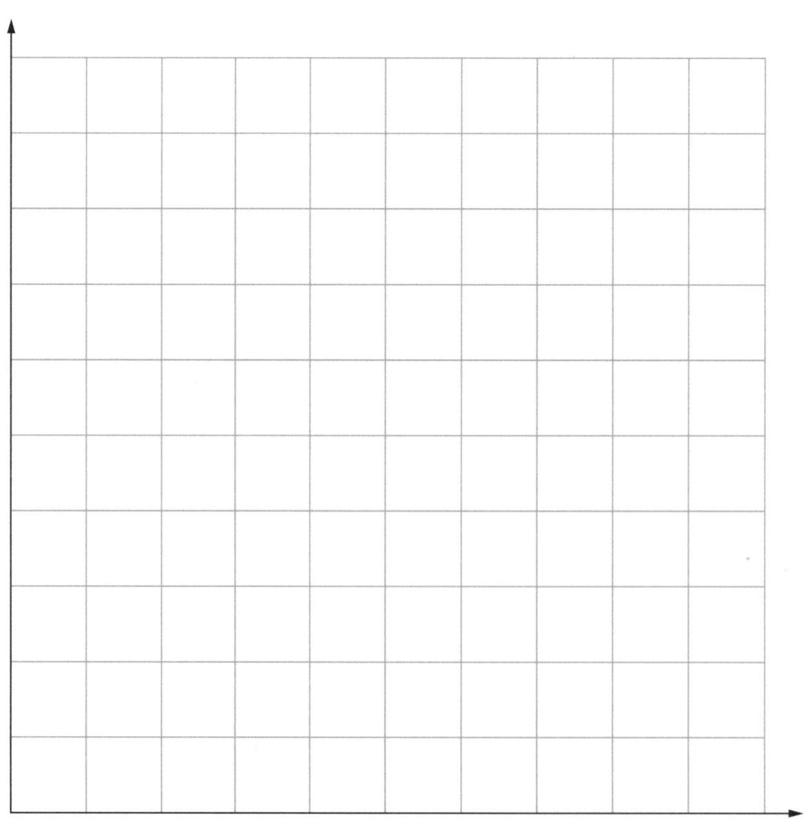

ISBN 978 1 4886 1936 6

3 Calculate the initial horizontal and vertical components of velocity for each angle, and fill in the columns in Table 2.

4 Use the horizontal distance and time of flight to calculate the average horizontal velocity for each angle, and fill in the corresponding column in Table 2. Why is this referred to as the *average* horizontal velocity?

5 Use the initial vertical velocity and the height of the launcher for each launch angle to calculate the theoretical time of flight of an object shot straight up. Record the results in Table 2.

TABLE 2 Projectile data calculations

Angle (°)	Initial horizontal velocity ($m\,s^{-1}$)	Initial vertical velocity ($m\,s^{-1}$)	Average horizontal velocity ($m\,s^{-1}$)	Theoretical time of flight (s)
0				
10				
20				
30				
40				
50				
60				
70				
80				

CONCLUSION

1 How did the measured horizontal velocities compare to the average horizontal velocities?

2 For any projectile launched horizontally, what can you state about the horizontal velocity?

3 Which launch angle will yield the maximum range (horizontal distance)?

4 Are there launch angles that yield the same range? What are they and why is that the case?

RATING MY LEARNING	My understanding improved	Not confident ◄——► Very confident ○ ○ ○ ○ ○	I answered questions without help	Not confident ◄——► Very confident ○ ○ ○ ○ ○	I corrected my errors without help	Not confident ◄——► Very confident ○ ○ ○ ○ ○

ISBN 978 1 4886 1936 6

PRACTICAL ACTIVITY 5.3

Circular motion—centripetal force in a horizontal plane

Suggested duration: 50 minutes

INTRODUCTION

The centripetal force, F_c, of an object of mass m moving at a constant velocity v and radius r is given by:

$$F_c = \frac{mv^2}{r} = 4\pi^2 rmf^2$$

where f is the frequency of its motion. The angular velocity, ω, of an object in a horizontal circular path is given by:

$$\omega = \frac{\Delta\theta}{t}$$

where θ is the angle subtended by the object in time t.

In this activity the centripetal force experienced by an object traveling in a horizontal plane will be investigated. A simple version of this experiment involves a mass being moved in a horizontal circular path. The addition of a force sensor to monitor the applied force rather than the use of a mass, or the use of a fully automated solution from one of the physics equipment manufacturers, can add a further degree of sophistication.

MATERIALS

- thin plastic tube about 15 cm long, with no sharp edges (the barrel of a ballpoint pen will do)
- 1.5 m of fishing line
- paperclip
- small soft mass (rubber stopper, cork or similar)
- mass carrier and slotted masses (50 g each)
- stopwatch
- metre rule

Substitute a force sensor for the mass carrier and slotted masses for real-time graphing of the centripetal force.

Warning: Students must wear safety glasses.

PURPOSE

To investigate the relationship between the centripetal force acting on an object moving in a circle of constant radius and the frequency of revolution.

PROCEDURE

1 Securely tie one end of the fishing line to the small, soft mass. (Since this is going to be twirled around your head, make sure the mass isn't too hard!)

2 Pass the fishing line down through the thin plastic tube and attach a 50 g slotted mass carrier to the end as shown in the diagram. Add three 50 g masses to the mass carrier to make a total mass of 200 g. If using slotted masses, secure them to the fishing line with sticky tape. Alternatively, connect the hook of a force sensor to the end of the fishing line instead of the mass carrier.

3 Attach a paperclip to the line to act as a marker for a measured radius of around 1 m. Measure and record the exact radius in the Data and analysis section.

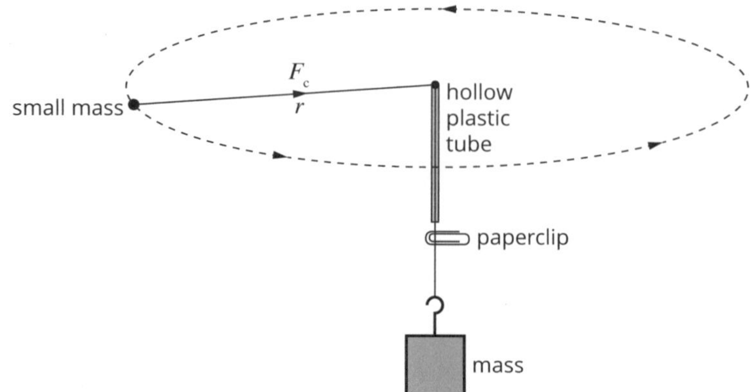

4 Twirl the stopper in a horizontal circular path at a speed that pulls the paperclip up to, but not touching, the bottom of the tube. Get a partner to keep an eye on the position of the clip to ensure that the speed of rotation stays quite constant. Practise doing this for a while before trying any measurements.

ISBN 978 1 4886 1936 6

5 Maintain the speed of revolution and measure the time taken for 20 revolutions of the small mass. Record this time in Table 1 of the Data and analysis section.

6 Add an extra 100 g to the mass carrier and repeat steps 2 and 3. Keep adding an extra 100 g mass to the mass carrier, until the mass carrier is full.

DATA AND ANALYSIS

Measured radius of revolution: __________ m

1 What force is the mass carrier providing in this experiment?

2 If using a force sensor, additional masses won't be added. What would you do instead to increase the applied force?

The force of gravity on the suspended mass is providing the centripetal force. Calculate the weight of the force acting on each mass using $g = 9.80\,\text{m}\,\text{s}^{-2}$. Alternatively, if using a force sensor, record the average recorded force for each test as the centripetal force. Complete Table 1 with the calculated results.

TABLE 1 Results

Mass (kg)	F_c **(N)**	**Time for 20 revolutions (s)**	**Period** T **for one revolution (s)**	$f = \frac{1}{T}$ **(Hz)**	**Angular velocity** ω **(° s^{-1})**

Plot a graph of frequency, f, versus centripetal force, F_c, and of frequency squared, f^2, versus centripetal force on the axes below.

PRACTICAL ACTIVITY 5.3

Frequency versus centripetal force

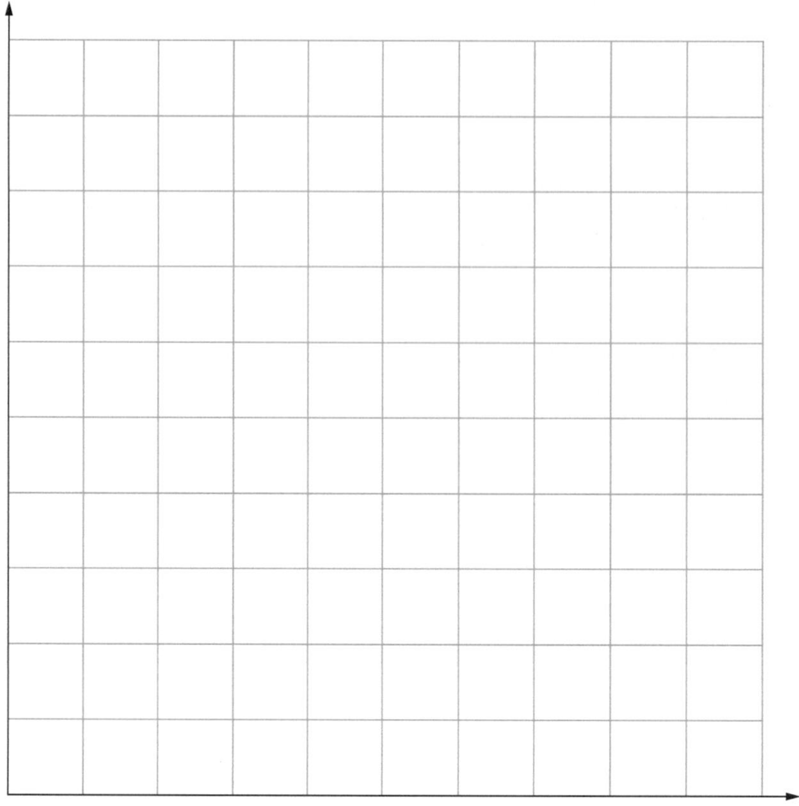

Frequency squared versus centripetal force

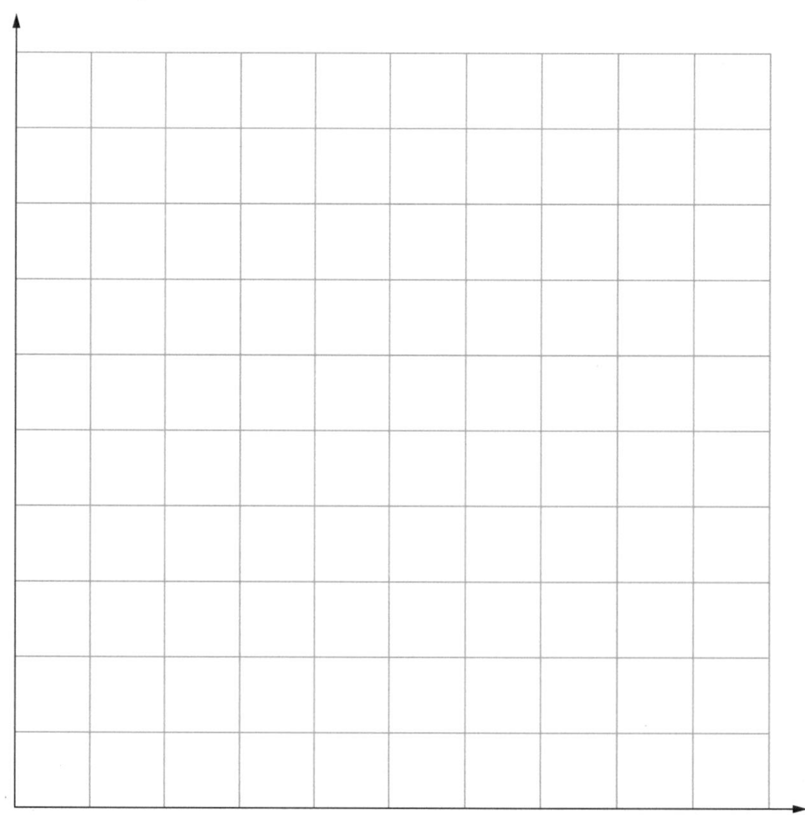

ISBN 978 1 4886 1936 6

3 What does the shape of the frequency squared versus centripetal force graph suggest?

4 What value should the gradient of this graph approximate?

CONCLUSION

1 Based on your results, what is the relationship between the centripetal force and the frequency of rotation? Do your results confirm what was expected from theory? Comment on any differences.

2 The radius of revolution will not actually be quite what was measured, nor will the tension in the string be exactly equal to the centripetal force. Why is this so?

3 What effect does this have on your results?

RATING MY LEARNING	My understanding improved	Not confident ◄──► Very confident ○ ○ ○ ○ ○	I answered questions without help	Not confident ◄──► Very confident ○ ○ ○ ○ ○	I corrected my errors without help	Not confident ◄──► Very confident ○ ○ ○ ○ ○

PRACTICAL ACTIVITY 5.4

Circular motion—centripetal force in a vertical plane

Suggested duration: 50 minutes

INTRODUCTION

In this activity, the centripetal force experienced by an object traveling in a vertical plane will be investigated.

PURPOSE

To investigate the non-uniform nature of the forces acting in vertical circular motion.

PROCEDURE

1 Tie the string to the mass and measure the length of the string to the centre of the mass. Mark the string in 20 cm segments.

2 To reduce the friction between the string and your fingers, fasten the alligator clip at one of the 20 cm marks. This will give a better defined radius of the mass and reduces friction, allowing a more constant speed of rotation and greatly improving accuracy.

3 Rotate the mass in a vertical circle. With practice, you can feel the change in tension in the string at the top of the circle compared with that at the bottom. Keep the rotation as even as possible.

4 Time the period of rotation for 20 revolutions and then find the average period. Repeat five times for each particular radius. You may need to use some additional paper if results are recorded for more than four radii.

5 Record your results in Table 1 of the Data and analysis section.

MATERIALS

- string
- small mass
- alligator clip
- stopwatch
- metre rule

As an option for more accurate timing, phones and tablets can be used to video the motion for video analysis.
Use a bright coloured mass against a plain background to allow for auto tracking through video analysis tools.

DATA AND ANALYSIS

TABLE 1 Radius and period

Radius, r (m)	Time for 20 revolutions (s)	Period T for one revolution (s)	T_{av} (s)

 ISBN 978 1 4886 1936 6

PRACTICAL ACTIVITY 5.4

Plot a graph of period versus radius on the axes provided.

Period versus radius

Why is a graph of period versus radius being plotted?

CONCLUSION

Based on your graph, comment on the relationship between period and radius. Do your results confirm what was expected from theory? Comment on any discrepancies.

RATING MY LEARNING	My understanding improved	Not confident ◄——► Very confident ○ ○ ○ ○ ○	I answered questions without help	Not confident ◄——► Very confident ○ ○ ○ ○ ○	I corrected my errors without help	Not confident ◄——► Very confident ○ ○ ○ ○ ○

DEPTH STUDY 5.1

Projectile motion—what happens when the wind blows?

Suggested duration: 3.75 hours

INTRODUCTION

Your initial studies of projectile motion mention air resistance as the other force, apart from gravity, acting on a projectile in flight, but for the purposes of the definition of projectile motion consider only the effect of gravity. In normal practice, of course, air resistance (including its variation with the projectile's velocity) and projectile spin play a big part in determining the final path of the projectile. Sports players can take advantage of spin to adjust the forces experienced by a ball in flight and to alter the flight of the ball. However, air resistance and the blowing of the wind can also have an unwanted effect that needs to be considered in setting the initial launch angle and speed.

This depth study requires you to question the role of the forces due to air resistance, spin and the wind on the flight path of a projectile. You will develop an inquiry question that requires research or experimentation to develop an informed hypothesis; plan an experimental investigation; and analyse primary- and secondary-sourced data and information from appropriate sources, using problem-solving techniques to determine the validity of the data and sources. You will evaluate the data to form conclusions by considering the quality of the data. You will process the data and information in order to communicate your findings in a poster format of approximately 500 words that includes your initial research or planning, experimentation, problem-solving and development of knowledge and understanding of the topic.

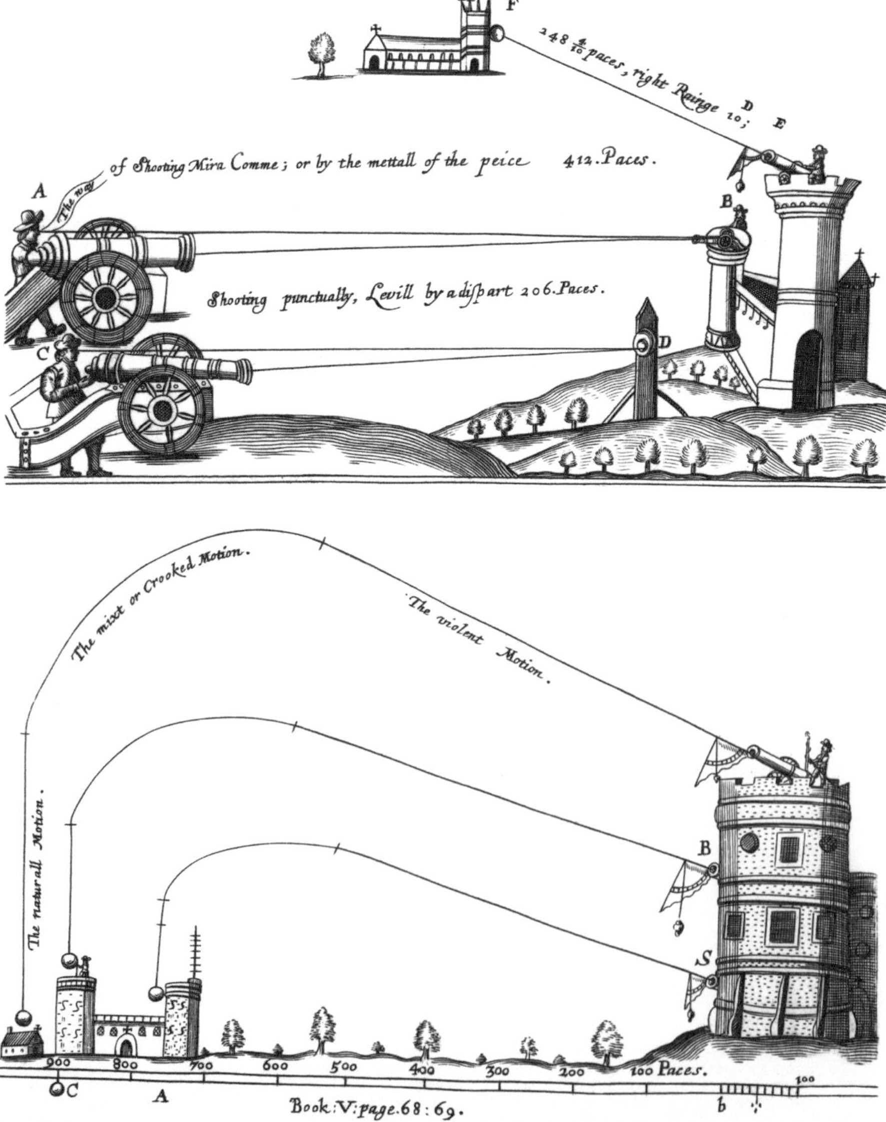

The potential path of a projectile has occupied the minds of scientists, mathematicians and others over many centuries. This illustration of the projected paths of cannonballs fired at different angles and heights was first published in *Mariners Magazine* in 1669.

ISBN 978 1 4886 1936 6

Your presentation will use appropriate scientific notation and nomenclature, graphs, force diagrams and appropriate calculations, and will appropriately apply scientific language that is suitable for the audience and context.

TOPIC REQUIREMENTS AND CONSTRAINTS

Your research must be conducted individually.

The topic chosen must relate to the motion of a projectile.

The topic must allow the development and answer of one clear inquiry question related to the effect of spin, wind, air resistance due to the shape of the object, or combinations of these, on the flight of the projectile, developing appropriate graphs of the flight path and associated vector diagrams.

PURPOSE

Research and experimentally investigate an application of projectile motion such as a particular ball sport, athletics event or 'human cannonball' phenomenon, to develop an appropriate hypothesis that explains the effect of wind, spin and other factors affecting air resistance on the projectile. Report on your knowledge and understanding of the topic based on an analysis of primary and secondary data.

QUESTIONING AND PREDICTING

1 What is a 'projectile' within the area of study of projectile motion?

2 Phrase your topic as a question. For example, 'How does backspin and top spin affect the path of a tennis ball in flight?' You may want to consult a few resources first. Choose a topic that is readily accessible and can be researched with the equipment that is available.

3 List three or four possible questions that would help answer your main question. For example, 'What force does backspin apply to a tennis ball?' or 'What force does top spin apply to a tennis ball?'

4 Do some preliminary research to help answer your initial questions. You may want to rephrase your questions to make them as clear and specific as possible. List your references and rephrased questions in the space provided.

5 Look back at your initial topic question. Construct a hypothesis that can be applied to answer your question. This becomes your working hypothesis and should summarise the answer to your main question. It may change after some further research.

PLANNING YOUR INVESTIGATION

6 Make note of any equations that are relevant to your investigation.

7 Describe your experimental method, justifying the reliability of the data you will collect. Start by creating a mathematical model of your projectile's flight, potentially using a spreadsheet program, and then follow this with a practical application of your model. Consider the use of video analysis to analyse the experimental flight. Keep the videos short and ensure that you are videoing from a position perpendicular to the motion so there is as little change in perspective as possible.

Comment on what steps you have taken to ensure the validity of the data as it applies to your hypothesis.

8 Working in small groups, evaluate other students' topics. What are the strengths of their experimental investigations or research? How could you improve your own investigation or research question?

 ISBN 978 1 4886 1936 6

CONDUCTING THE INVESTIGATION

9 Refine your mathematical model by synthesising your research from secondary sources.

Carry out the remainder of your investigation, summarising key points as you proceed so that your topic has a clear focus and development. Use the tree diagram below to link your development back to your original hypothesis.

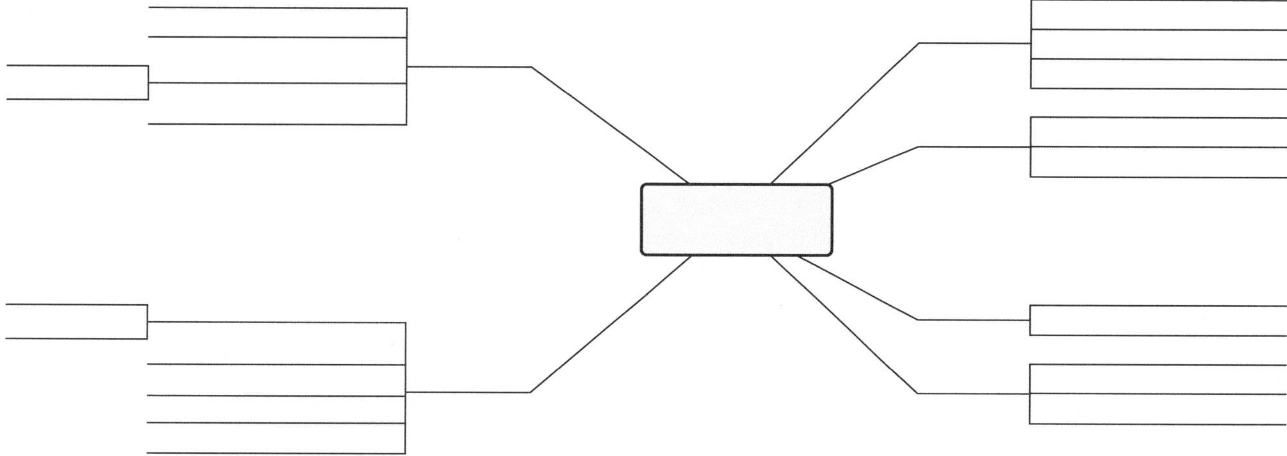

ANALYSING DATA AND INFORMATION

10 Were you successful in answering your original hypothesis? Did you need to rephrase it?

11 Does your experimental data support your conclusions? How reliable is your primary and secondary data? You may be able to calculate a reasonable measure of the likely error in your results.

PROBLEM SOLVING

12 To what extent do your findings confirm or contradict other sources?

13 How well did your mathematical model fit your experimental results? You may want to calculate a percentage difference between the experimental and theoretical results.

ISBN 978 1 4886 1936 6

14 Comment on any assumptions made within your model and whether you accounted for them within your experiment.

COMMUNICATING

Communicate your findings in the form of a scientific poster. It should include explanations of the topic development and/or experimental methodology, and assumptions you have made, relevant calculations, data, graphs and clear vector diagrams illustrating the effect of the additional outside force/s on the flight of the projectile. A suitable template for your poster is below.

Title (*inquiry question, ending in a '?'*)
Student name
School name
Year 12 Physics

Introduction

Cut and paste from Microsoft® Word® document/logbook

- *Include purpose and hypothesis.*
- *Summarise background physics concepts, including in-text citations to appropriate references.*
- *Font should be easy to read from 1 m away (suggest minimum 16 point in Calibri, Arial or Tahoma).*
- *Use contrasting colours (e.g. black or blue colour on white background).*

Cut and paste images (e.g. from logbook)

Where possible include a labelled diagram of experimental set-up/results

Each figure should be numbered with an appropriate label (and reference where applicable)

Analysis

Cut and paste from Word document/ logbook

- *Discuss, analyse and evaluate data obtained (directly refer to data in tables).*
- *Link results to key physics concepts (include in-text citations as appropriate).*
- *Identify outliers.*
- *Include limitations and recommendations for further research.*

Data

Cut and paste from Word document/logbook

- *Include summary of results obtained.*
- *Avoid including all raw and processed experimental data—use extract of raw and processed experimental data.*
- *Where possible, present data graphically (e.g. in pie/bar/line graphs).*
- *Number each table/graph and give it an appropriate title.*
- *Ensure axes are labelled, including units if applicable.*

Procedure

Cut and paste from Word document/ logbook

- *Include summary of method used:*
  - *e.g. flow chart*
  - *consider using embedded graphics in media presentation programs such as Microsoft® PowerPoint®.*
- *Identify and manage relevant risks and follow relevant safety guidelines.*

Conclusion

Cut and paste from Word document/ logbook

- *Summarise research findings according to aim and hypothesis.*
- *Include recommendations for future research.*

References

Include alphabetical listing (by author last name/institution) of all sourced material; ensure in-text citations included where appropriate in poster

Use consistent referencing style (e.g. APA)

Acknowledgements

 ISBN 978 1 4886 1936 6

MODULE 5 • REVIEW QUESTIONS

Multiple choice

1 The figure shows a view from above a hammer thrower at the instant of release. The arrow shows the direction of rotation of the athlete and hammer.

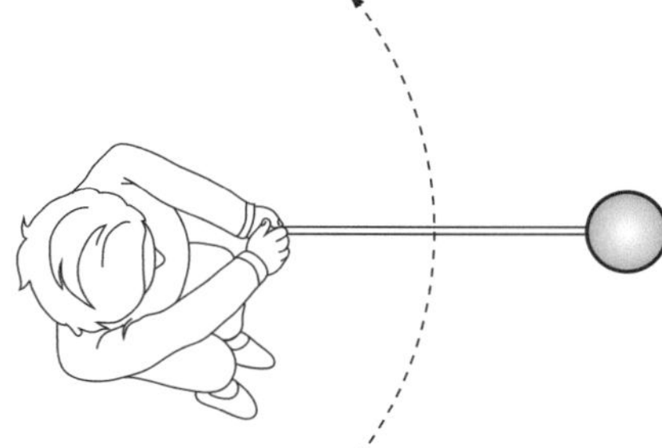

Which one of the diagrams A to E below best shows the path of the ball after release?

A 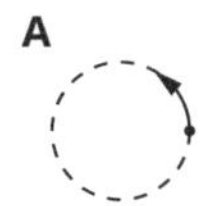  **B** 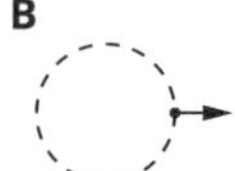 **C**  **D** 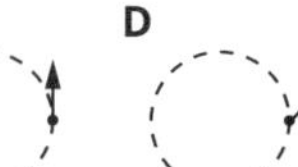 **E** 

2 Which one or more of the following are correct? From a location on the Earth's equator, the apparent position of a geostationary satellite:

A reappears directly overhead every 24 hours.

B moves between two positions north and south of the location.

C remains directly overhead.

D remains directly overhead but only during the day.

3 Which one or more of the following are correct? The quantity $\frac{r^3}{T^2}$ is the same for all the:

A planets in the solar system.

B planets and moons in the solar system.

C planets and asteroids in the solar system.

D planets excluding Pluto.

4 In the absence of air resistance, the maximum range of a projectile occurs when the angle of elevation is:

A 30°

B 45°

C 60°

D 90°

5 Simple manipulation of Newton's equation $F = -\frac{GMm}{r^2}$ gives an expression for g with a value of $9.80\,\text{m}\,\text{s}^{-2}$. Which one of the descriptions below does not describe the term g?

A g is the gravitational field at the surface of the Earth.

B g is the force that a mass m feels at the surface of the Earth.

C g is the force experienced by a mass of 1 kg at the surface of the Earth.

D g is the acceleration of a free body at the surface of the Earth.

6 The planet Mercury has a mass about $\frac{1}{20}$ of the Earth's mass and a radius of about $\frac{2}{5}$ Earth's radius. What is the approximate gravitational field on Mercury's surface?

A $0.40g$

B $0.05g$

C $0.13g$

D $0.31g$

Short answer

7 A golfer hits a golf ball at $33.0\,\text{m}\,\text{s}^{-1}$ at an angle of 26.0° to the horizontal and it lands on the green, 5.00 m below her position. She was trying to stop the ball close to the flag, which is 107 m away.

a Calculate the time that the ball is in the air before hitting the green and then find how far horizontally from the golfer the ball first strikes the green.

b What is the velocity of the ball as it lands?

c Why would a higher shot be preferred if the golfer wanted to stop the ball as near to the hole as possible?

8 A 0.40 kg object is whirled in a horizontal circle (as a conical pendulum) on the end of a string 1.80 m long. The string makes an angle of 23° to the vertical.

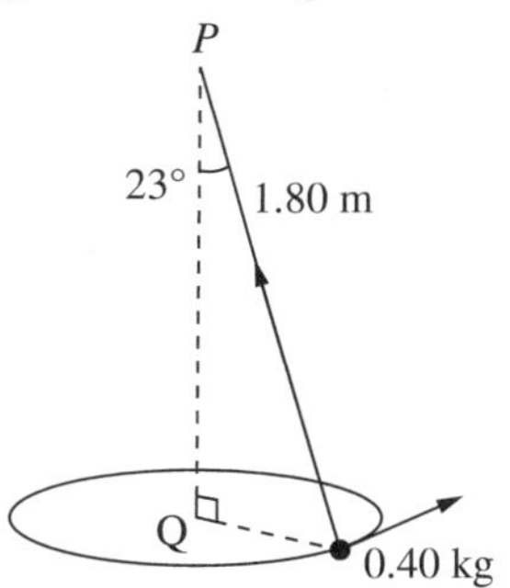

a What is the net vertical force acting on the object?

b What is the radius of the object's path?

c Calculate the tension in the string.

d How fast is the object travelling?

9 A catapult is set up to throw a rock over a 16.0 m high castle wall 65 m away. The initial velocity and angle of elevation are 48 m s^{-1} and 32° respectively. The catapult and the base of the wall are on the same horizontal level.

a How long will it take for the rock to pass over the wall?

b By what distance will the rock clear the wall?

10 A small space probe is placed in a very low orbit around the Moon (effectively at an altitude of 0 m).

a Use the equations of gravity and circular motion to show that the Moon's average density can be expressed as $\frac{3\pi}{GT^2}$.

b If the orbital period of the probe is observed to be 109 minutes, calculate the Moon's average density.

11 A rocket-powered spacecraft of mass 125 kg (including fuel) is in Earth's orbit at an altitude of 6.30×10^2 km.

a Show that the speed v of the satellite is given by $v^2 = \frac{GM_e}{r}$.

b Calculate the speed and kinetic energy of the satellite.

The controllers at NASA want to alter the orbit to a higher path at 730 km above the Earth's surface. They do this by firing the satellite's thruster engine for a short time, in the direction of the satellite's velocity.

c What will be the satellite's kinetic energy in the new orbit?

d Explain how an initial increase in the satellite's speed ends in reducing its kinetic energy.

 ISBN 978 1 4886 1936 6

MODULE 6

Electromagnetism

Outcomes

By the end of this module you will be able to:

- develop and evaluate questions and hypotheses for scientific investigation PH12-1
- design and evaluate investigations in order to obtain primary and secondary data and information PH12-2
- conduct investigations to collect valid and reliable primary and secondary data and information PH12-3
- select and process appropriate qualitative and quantitative data and information using a range of appropriate media PH12-4
- analyse and evaluate primary and secondary data and information PH12-5
- explain and analyse the electric and magnetic interactions due to charged particles and currents and evaluate their effect both qualitatively and quantitatively. PH12-13

Content

CHARGED PARTICLES, CONDUCTORS AND ELECTRIC AND MAGNETIC FIELDS

INQUIRY QUESTION **What happens to stationary and moving charged particles when they interact with an electric or magnetic field?**

By the end of this module you will be able to:

- investigate and quantitatively derive and analyse the interaction between charged particles and uniform electric fields, including: (ACSPH083) ICT N
  - electric field between parallel charged plates $E = \frac{V}{d}$
  - acceleration of charged particles by the electric field $\vec{F}_{net} = m\vec{a}$, $\vec{F} = q\vec{E}$
  - work done on the charge $W = qV$, $W = qEd$, $K = \frac{1}{2}mv^2$
- model qualitatively and quantitatively the trajectories of charged particles in electric fields and compare them with the trajectories of projectiles in a gravitational field CCT ICT L N
- analyse the interaction between charged particles and uniform magnetic fields, including: (ACSPH083)
  - acceleration, perpendicular to the field, of charged particles
  - the force on the charge $F = qv_{\perp}B = qvB\sin\theta$ N
- compare the interaction of charged particles moving in magnetic fields to: CCT
  - the interaction of charged particles with electric fields
  - other examples of uniform circular motion (ACSPH108).

Module 6 • Electromagnetism

THE MOTOR EFFECT

INQUIRY QUESTION **Under what circumstances is a force produced on a current-carrying conductor in a magnetic field?**

By the end of this module you will be able to:

- investigate qualitatively and quantitatively the interaction between a current-carrying conductor and a uniform magnetic field $F = lI_{\perp}B = lIB\sin\theta$ to establish: (ACSPH080, ACSPH081) CCT ICT N
  - conditions under which the maximum force is produced
  - the relationship between the directions of the force, magnetic field strength and current
  - conditions under which no force is produced on the conductor
- conduct a quantitative investigation to demonstrate the interaction between two parallel current-carrying wires
- analyse the interaction between two parallel current-carrying wires $\frac{F}{l} = \frac{\mu_0}{2\pi} \times \frac{I_1 I_2}{r}$ and determine the relationship between the International System of Units (SI) definition of an ampere and Newton's third law of motion (ACSPH081, ACSPH106). ICT L N

ELECTROMAGNETIC INDUCTION

INQUIRY QUESTION **How are electric and magnetic fields related?**

By the end of this module you will be able to:

- describe how magnetic flux can change, with reference to the relationship $\Phi = B_{\parallel}A = BA\cos\theta$ (ACSPH083, ACSPH107, ACSPH109) ICT N
- analyse qualitatively and quantitatively, with reference to energy transfers and transformations, examples of Faraday's law and Lenz's law $\varepsilon = -N\frac{\Delta\Phi}{\Delta t}$, including but not limited to: (ACSPH081, ACSPH110) ICT N
  - the generation of an electromotive force (emf) and evidence for Lenz's law produced by the relative movement between a magnet, straight conductors, metal plates and solenoids
  - the generation of an emf produced by the relative movement or changes in current in one solenoid in the vicinity of another solenoid
- analyse quantitatively the operation of ideal transformers through the application of: (ACSPH110) ICT N
  - $\frac{V_p}{V_s} = \frac{N_p}{N_s}$
  - $V_p I_p = V_s I_s$
- evaluate qualitatively the limitations of the ideal transformer model and the strategies used to improve transformer efficiency, including but not limited to: CCT
  - incomplete flux linkage
  - resistive heat production and eddy currents
- analyse applications of step-up and step-down transformers, including but not limited to:
  - the distribution of energy using high-voltage transmission lines. CCT

APPLICATIONS OF THE MOTOR EFFECT

INQUIRY QUESTION **How has knowledge about the motor effect been applied to technological advances?**

By the end of this module you will be able to:

- investigate the operation of a simple DC motor to analyse:
  - the functions of its components
  - production of a torque $\tau = nIA_{\perp}B = nIAB\sin\theta$
  - effects of back emf (ACSPH108) CCT ICT N
- analyse the operation of simple DC and AC generators and AC induction motors (ACSPH110) CCT
- relate Lenz's law to the law of conservation of energy and apply the law of conservation of energy to:
  - DC motors, and
  - magnetic braking. CCT

Key knowledge

Charged particles, conductors, and electric and magnetic fields

PARTICLES IN ELECTRIC FIELDS

An **electric field** is a region of space around a charged object in which another charged object will experience a force. An electric field has both strength and direction, which makes it a vector quantity. Electric fields are represented using field lines. Electric field lines point in the direction of the force that a positive charge within the field would experience. A positive charge experiences a force in the direction of the electric field and a negative charge experiences a force in the opposite direction to the field. The spacing between the field lines indicates the strength of the field. The closer together the lines are, the stronger the field. Figure 6.1 shows an electric field between a positive and negative charge, while also indicating the strength of the field.

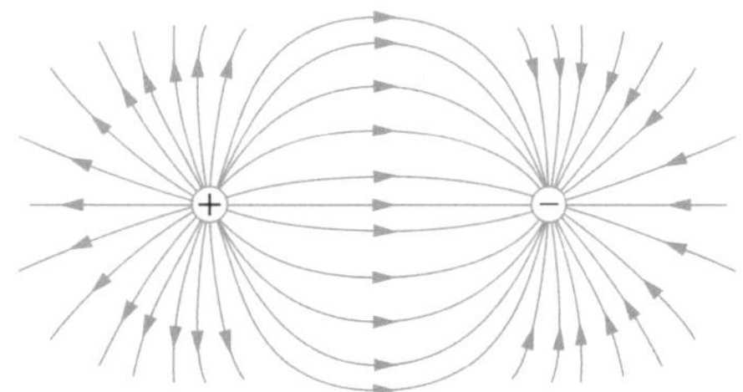

FIGURE 6.1 A positive and a negative charge with the field lines between them

Electric field strength can be expressed as:

$$\vec{E} = \frac{\vec{F}}{q}$$

where:

$\vec{E}$ is the strength of the electric field (N C^{-1})

$\vec{F}$ is the force on the charged particle (N)

q is the charge of the object experiencing the force (C).

It is important to note that in calculating the electric field strength, q represents the charge of the object experiencing the force, also known as a **point charge**. This point charge allows the electric force to be measured, but is not large enough to create a significant force on any other charges. Around point charges the electric field radiates in all directions (three dimensionally).

Charges in an electric field will accelerate in the direction of the force acting on them. This acceleration can be determined by relating the following equations:

$$\vec{F} = q\vec{E} \text{ and } \vec{F}_{\text{net}} = m\vec{a}$$

where:

m is the mass of the accelerating particle (kg)

$\vec{a}$ is the acceleration (m s^{-2}).

If the particle increases its velocity, decreases its velocity, or changes its direction while in the field, then the charged particle is undergoing acceleration in the field.

Electrical potential (V) is defined as the work required per unit charge to move a positive point charge from infinity to a point within the electric field. The electrical potential at infinity is defined as zero.

The potential across a distance d and the **electric field strength** E are related by the following equation:

$$E = \frac{V}{d}$$

Between two oppositely charged parallel plates (two points), the field lines are parallel and therefore the field has a uniform strength. Figure 6.2 represents a uniform electric field between two parallel plates. The potential difference is the change (shown by the delta symbol, Δ) between these two points. The electric potential however, is the actual work required per charge to move a positive point charge from zero to the positively charged plate.

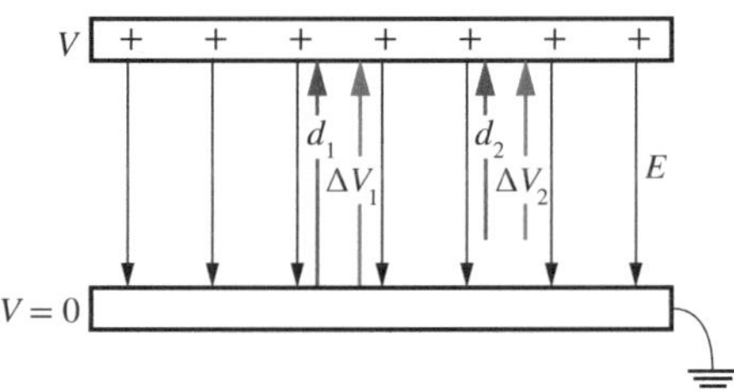

FIGURE 6.2 The potential difference is the change in potential (ΔV) between two points a distance d apart in a uniform electric field.

Electric potential energy is the energy a charge has due to its position relative to other charges. It is a form of energy that is stored in an electric field. Work is done on the field when a charged particle is forced to move in the electric field. Conversely, when energy is stored in the electric field then work can be done by the field on the charged particle.

Work is done whenever a force moves something over a distance. Work is a measure of the amount of energy used in moving the object.

When a charged object is moved against the direction it would naturally move in an electric field, then work is done on the field. When a charged object moves in the direction it would naturally tend to move in an electric field, then the field does work on the particle. For example, Figure 6.3 shows the motion of three electrons in an electric field. An electron will naturally move towards the positive plate, so for the motion of q_1 work is done by the field, while for q_2 work is done on the field. If the charge doesn't move any distance parallel to the field then no work is done, i.e. q_3 moves perpendicular to the field so no work is done.

 ISBN 978 1 4886 1936 6

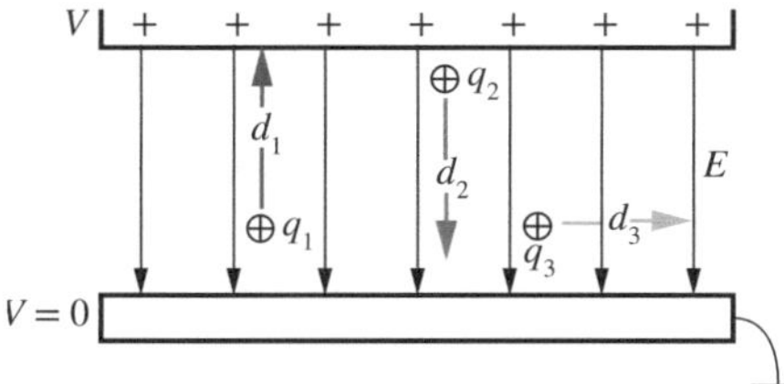

FIGURE 6.3 Work is being done on the field by moving q_1 and work is being done by the field on q_2. No work is done on q_3 since it is moving perpendicular to the field.

The work done on or by an electric field can be calculated using the equations:

$$W = qV \text{ or } W = qEd$$

where:

W is the work done on the point charge or on the field (J)

V is the potential difference between points (V)

q is the charge of the point charge (C)

E is the electrical field strength (V m^{-1} or N C^{-1})

d is the distance between points, parallel to the field (m).

If the electric force created by the field on the charged object is equal to the gravitational force on (or weight of) the object, then these two forces can add to provide a net force equal to zero. This means that the charged object will either be suspended between the plates, or (by Newton's first law of motion) will be falling or rising at constant velocity.

Gravitational, electric and magnetic fields are similar, but display significant differences associated with the differences in the fundamental nature of the fields. The direction of a field at any point is always the resultant field vector determined by adding the individual field vectors due to each mass (gravitational field), charge (electric field) or magnetic pole (magnetic field) within the affected region. In a static (unchanging) field, the strength of the field doesn't change with time.

Just as projectiles have parabolic motion as they are only acted on by gravity, charged particles moving within electric fields can have a similar shape. So long as the initial conditions of the particle are known (i.e. its initial velocity), the motion of the particle can be analysed using the equations of motion, similar to modelling projectiles in Module 5.

PARTICLES IN MAGNETIC FIELDS

A **magnetic field** is generated from the magnetic effect of electric currents and magnetic materials. Like an electric field, a magnetic field is a vector quantity and can be illustrated in the same way as an electric field; however, it always contains two poles.

GO TO ➤ Year 11 Module 4

When a charged particle travels into a magnetic field, the force on it is perpendicular to both its direction of motion and the magnetic field lines; in comparison a charged particle in an electric field will travel in the direction of the electric field lines.

The magnitude of the force on a charged object within a magnetic field is proportional to the strength of the magnetic field, B, the velocity of the charge, the angle the object is moving at with respect to the magnetic field, and the charge on the particle. This is represented by:

$$F = qv_{\perp}B = qvB\sin\theta$$

where:

F is the force (N)

q is the electric charge on the particle (C)

v is the instantaneous velocity of the particle (m s^{-1})

B is the strength of the magnetic field (T)

θ is the angle the object is moving at with respect to the magnetic field (°).

This force is referred to as the Lorentz force. The force is at a maximum when the charged particle is moving at right angles to the magnetic field (shown by the perpendicular symbol ⊥). The force is zero when the charged particle is travelling parallel to the magnetic field.

The right-hand rule: Point the thumb of the right hand in the direction of the movement of *a positive charge* (conventional current direction) and the fingers in the direction of the magnetic field. The force on the charge will point out from the palm.

The right-hand rule is used to determine the direction of the force on a positive charge moving in a magnetic field, B. The direction of the force on a negatively charged particle is in the opposite direction.

If a moving charge experiences a force of constant magnitude that remains at right angles to its motion, its direction will be changed but not its speed. The result is a circular orbit, shown in Figure 6.4.

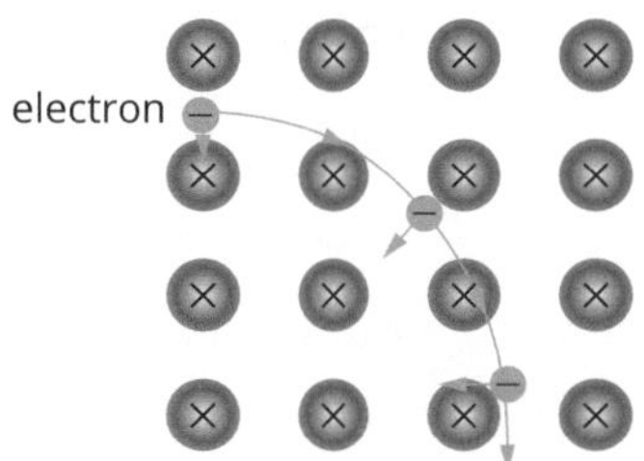

FIGURE 6.4 The crosses illustrate the magnetic field going into the page and the charged particle moving at right angles to it.

The consequences of the circular motion of charged particles can be used for a range of practical applications. Many technologies are based on the motion of charged particles in electromagnetic fields. Particle accelerators are machines that accelerate charged particles, such as electrons, protons or atomic nuclei, to speeds close to that of light. The particles are accelerated by electromagnetic fields, but very long paths are required for the particles to obtain the extremely high speeds needed. To achieve this without the need for tunnels hundreds of kilometres long, particles travel through very strong magnetic fields that cause them to move in a circle. As they accelerate around curves, the electrons give off bursts of radiation which are utilised by researchers in a range of experimental stations.

The motor effect

FORCE ON A CONDUCTOR

A magnetic field is produced around a flow of charge (i.e. a current), and a force is experienced by a charge within this field. The magnitude of the force on a charged object within a magnetic field is proportional to the strength of the magnetic field, B, the component of the velocity of the charge that is perpendicular (at right angles) to the magnetic field, and the charge on the particle, i.e. $F = qvB$.

A conducting wire allows a stream of charged particles to flow in one direction. A flow of charged particles is a **current**. Electron flow is different to conventional current. Conventional current I is defined as moving in the direction a positive charge would travel. If a conductor carrying many charged particles is placed within a magnetic field, it will experience a force. This is the theory behind the operation of electric motors. The force acting on any conductor, or part of a conductor, moving at an angle θ to the magnetic field, is given by the equation

$$F = lI_{\perp}B = lIB\sin\theta$$

where:

F is the force on the conductor (N)

l is the length of the conductor (m).

I is the current in the conductor (A)

B is the strength of the magnetic field (T)

This equation can also be applied to situations where there are multiple coils n. The force is then found using $F = nlI_{\perp}B$.

The force is at a maximum when the charged particle is moving at right angles to the magnetic field. The force is zero when the charged particle is travelling parallel to the magnetic field. The right-hand rule can be used to determine the direction of the force on the charge.

Electric motors are built to ensure that the current-carrying conductor is perpendicular to the magnetic field, so that the motor utilises the full effect of the magnetic force. Consider Figure 6.5, a single square coil of wire, with vertices ABCD, carrying a current I in a magnetic field B. Initially the wire coil is aligned horizontally in a magnetic field, as in Figure 6.5a. Sides AD and BC are parallel to the magnetic field so no magnetic force will act on them. Sides AB and CD are perpendicular to the field so both of these sides will experience a magnetic force. Using the right-hand rule, there is a downwards force on AB and an upwards force on CD. These two forces will act together on the coil and cause it to rotate anticlockwise (Figure 6.5b). As the coil rotates to the position shown in Figure 6.5c, the forces acting on each side are such that they will tend to keep the coil in this position. For the coil to continue to rotate anticlockwise at this point, the current direction needs to be reversed. This is shown in Figure 6.5d.

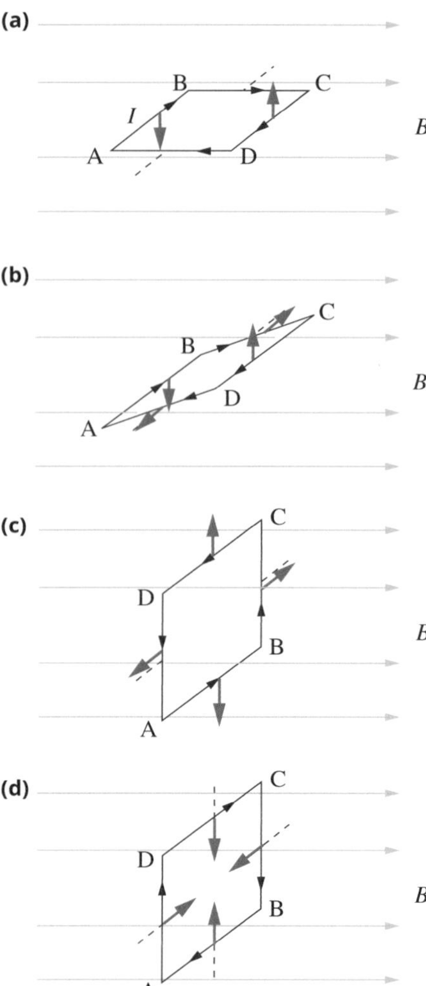

FIGURE 6.5 The magnetic force acting on each side of a current-carrying square wire coil as it rotates from being (a) parallel to the field through to (d) perpendicular to the field.

FORCES BETWEEN CONDUCTORS

When two conductors are placed parallel to each other, an interesting effect occurs. As discussed above, if a current flows, it will have an associated magnetic field. Depending on the direction of the current in the conductors (same direction or opposite) the resulting magnetic forces will either attract or repel.

The force, F, between two parallel current-carrying conductors is given by the following equation:

$$F = I_1 I_2 \frac{\mu_0}{2\pi r} l$$

where:

I_1 and I_2 are the currents in the parallel conductors (A)

μ_0 is the magnetic permeability of free space ($4\pi \times 10^{-7}\,\text{NA}^{-2}$)

r is the radius from the current-carrying conductor to the location where the magnetic field is measured (m)

l is the length of the conductors (m).

This equation can be re-arranged to show the force per unit length:

$$\frac{F}{l} = \frac{\mu_0 I_1 I_2}{2\pi r}$$

 ISBN 978 1 4886 1936 6

If the two parallel conductors carry current in the same direction, the forces attract. If the two parallel conductors carry current in opposite directions, the forces will repel. This is illustrated in Figure 6.6. The force experienced between the two parallel conductors is equal and opposite—an example of Newton's third law; i.e. the force wire 1 exerts on wire 2 is equal and opposite to the force wire 2 exerts on wire 1.

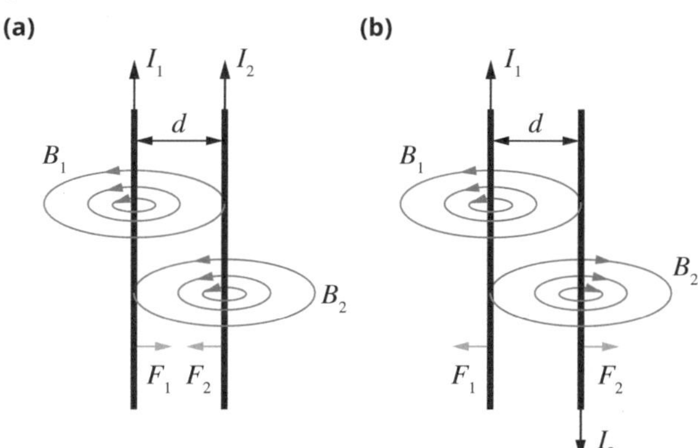

FIGURE 6.6 Two parallel conductors, with (a) current going in in the same direction and (b) current going in opposite directions

The direction of the magnetic field around each conductor can be found using the right-hand grip rule. Point your thumb in the direction of conventional current; then your curled fingers will point in the direction of the magnetic field.

The forces between two parallel current-carrying conductors are used to derive the fundamental definition of the ampere. If the conductors are 1 m apart, and each conductor carries 1 A of current, this equation can be simplified to:

$$\frac{F}{l} = \frac{\mu_0 \times (1\text{ A}) \times (1\text{ A})}{2\pi \times (1\text{ m})}$$
$$= \frac{\mu_0}{2\pi}$$
$$= \frac{4\pi \times 10^{-7}}{2\pi}$$
$$= 2 \times 10^{-7}\text{ N m}^{-1}$$

This means that one ampere of current through two infinitely long parallel conductors (separated by one metre in empty space and free of any other magnetic fields) causes a force of $2 \times 10^{-7}\text{ N m}^{-1}$ on each conductor.

Electromagnetic induction

MAGNETIC FLUX

In the 19th century, scientists studying the relatively new field of electrical currents discovered that moving charges produce magnetic effects. After this, Michael Faraday, an English scientist, was convinced that the reverse should also be true—a magnetic field should be able to produce an electric current. It was discovered that a *change* in a magnetic field, when a magnet is moved closer to a conductor, leads to an induced **emf** (electromotive force) that in turn produces a current. A way to induce an emf is by moving a magnet in and out of a coil or moving a straight conductor in a magnetic field.

In Figure 6.7, the charges in the wire are moving apart due to the force they are experiencing from the magnetic field. One end of the conductor will become more positively charged, the other more negatively charged and a potential difference, V, or emf will be induced between the ends of the conductor. This can be represented by the formula:

$$\varepsilon = lvB$$

where:

ε is the induced emf (V)

l is the length of the conductor (m)

v is the speed of the conductor perpendicular to the magnetic field (m s^{-1})

B is the strength of the magnetic field (T).

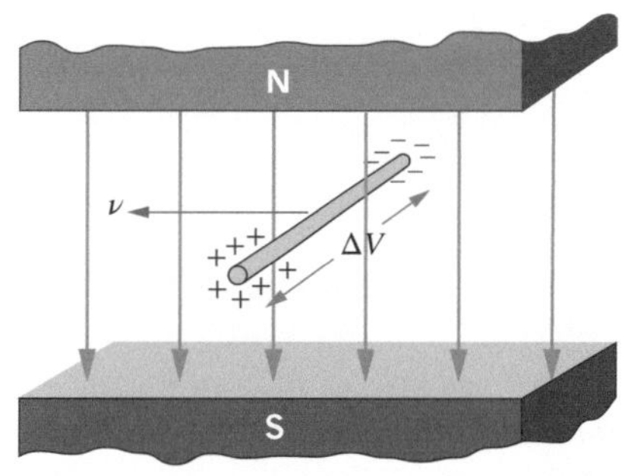

FIGURE 6.7 A potential difference, ΔV, will be produced across a straight wire moving to the left in a downward-pointing magnetic field.

The production of an induced emf, ε, by a changing magnetic flux is called **electromagnetic induction**. **Magnetic flux** is essentially the total number of magnetic field lines passing through any given area. Figure 6.8a represents a strong magnetic field acting over a small area, which has the same magnetic flux as a weaker magnetic field acting over a larger area (Figure 6.8b).

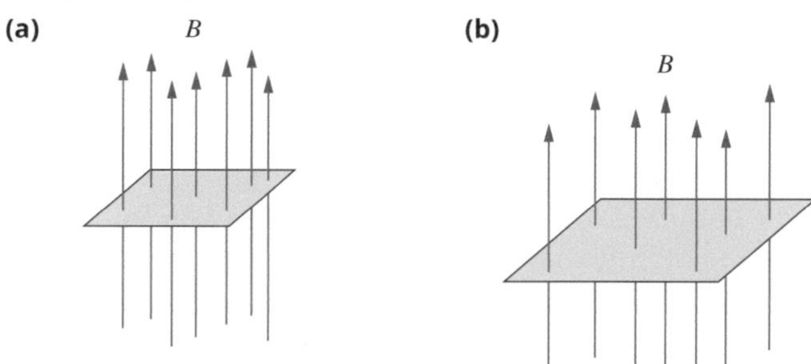

FIGURE 6.8 The magnetic flux is the same in (a) and (b).

Magnetic flux is defined as the product of the strength of the magnetic field B and the area of the field perpendicular to the field lines, A. It is given by the formula:

$$\Phi = B_{\parallel}A$$

where:

Φ is the magnetic flux in weber (Wb, where $1\text{ Wb} = 1\text{ T m}^2$)

$B_{\parallel}$ is the magnetic field (T)

A is the area vector (m^2).

The area being examined is actually represented by a vector A, the magnitude of which is equal to the area being examined. The direction of the area vector is normal to the plane of the area. The magnetic flux will be at a maximum when the area vector is parallel to the magnetic field. The subscript || is included in the formula to indicate that the magnetic field is parallel to the area vector.

By changing the angle between the magnetic field and the normal to the area (the area vector), the magnetic flux varies. When the normal to the area is parallel to the field (at 0°), the flux is a maximum. When the normal to the area is perpendicular to the field (at 90°), the flux is zero. Be careful in your calculation not to confuse the angle between the plane of the area and the magnetic field, and the angle between normal to the area and the magnetic field. Drawing a diagram when you complete a magnetic flux problem can be helpful. This can be represented by the following formula:

$$\Phi = B_{\parallel}A = BA\cos\theta$$

The total flux of multiple coils within a magnetic field is sometimes referred to as the flux linkage.

Eddy currents are an example of induced electric currents. For example, if a metal plate is dragged out of a magnetic field, a circular eddy current will form within the plate that opposes the change in flux through the area of the plate, and thus opposes the motion of the plate itself due to the interaction of the magnetic fields.

FARADAY'S AND LENZ'S LAWS

Faraday's investigations led him to conclude that the average emf induced in a conducting loop, in which there is a changing magnetic flux, is proportional to the negative rate of change of flux. This is now known as **Faraday's law** of induction and is one of the basic laws of electromagnetism. If the flux through N turns (or loops) of a coil changes from Φ_1 to Φ_2 during a time t, then the average induced emf during this time will be:

$$\varepsilon = -N\frac{\Delta\Phi}{\Delta t}$$

The negative sign in Faraday's law indicates direction. For questions involving only magnitudes, you should not use the negative sign or any negative quantities.

The Russian physicist Henrich Lenz developed a law to understand how electromagnetic induction obeys the principles of conservation of energy and which explains the direction of the induced emf. This is known as **Lenz's law**. He used the idea of nature trying to oppose any applied force. Lenz's law states that when an emf is generated by a change in magnetic flux (Faraday's law), the induced emf produces a current which in turn produces a magnetic field that opposes the original change in flux. For example, when an induced current flows through a coil and the **galvanometer**, the magnetic field produced by the coil has a direction that repels the incoming permanent magnet pole. The same happens if the permanent magnet is pulled out of the coil; however, the magnetic field produced tries to attract the withdrawing permanent magnet back inside the coil again.

> A galvanometer is a type of sensitive ammeter: an instrument for detecting a small electric current.

In Figure 6.9, the north end of a magnet is brought towards a coil from right to left, inducing a magnetic field where the current flows anticlockwise.

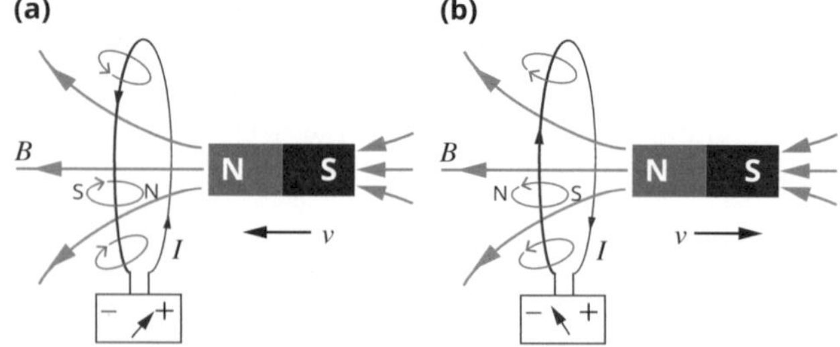

FIGURE 6.9 (a) The north end of a magnet is brought towards a coil from right to left, inducing a current that flows anticlockwise. (b) Pulling the north end of the magnet away from the coil from left to right induces a current in a clockwise direction.

There are three distinct steps to determine the induced current direction according to Lenz's law:

1. What is the change that is happening?
2. What will oppose the change and/or restore the original conditions?
3. What must be the current direction to match this opposition?

For example, consider Figure 6.10, where the arrow indicates the direction the magnet is moving. In this figure the south pole of a magnet is moved towards the horizontal coil held about it.

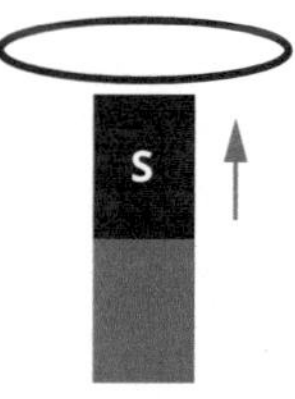

FIGURE 6.10 The south end of a magnet is brought towards a coil.

In order to find the direction of the induced current, the three distinct steps will be followed.

1. What is the change that is happening?

The magnetic field direction in the coil will be downwards towards the south pole of the magnet. As the magnet is brought closer to the coil, the downwards flux from the coil will increase from the magnet (the coil initially had zero magnetic field, hence zero magnetic flux). So the change in flux is increasing downwards.

2. What will oppose this change?

As the magnetic flux is increasing downwards as the magnet is moving towards the coil, this means the induced magnetic field that opposes the change would act upwards.

3. What is the current direction?

In order to oppose the change, the current direction would be anticlockwise when viewed from above (using the right-hand grip rule).

 ISBN 978 1 4886 1936 6

It is important to note it is the flux, not the field, and the rate of change of flux, not the absolute size of the flux that matters. An induced emf can be created in three ways:

- by changing the strength of the magnetic field
- by changing the area of the coil within the magnetic field
- by changing the orientation of the coil with respect to the direction of the magnetic field.

TRANSFORMERS

A transformer is a device for increasing and decreasing an alternating current (AC) voltage, which is a very essential part of any electrical distribution system. A transformer works on the principle of a changing magnetic flux inducing an emf. No matter what the size or application, it will consist of two coils known as the primary and secondary coils. These two coils are wound onto a common soft iron core, shown in Figure 6.11a.

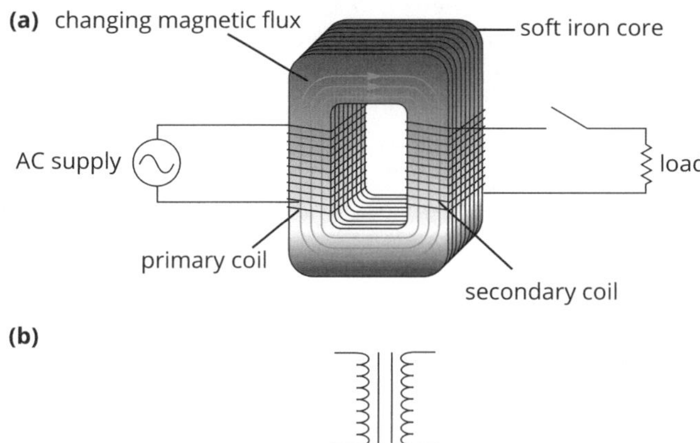

FIGURE 6.11 (a) A transformer and (b) the symbol used in circuit diagrams for transformers

If the primary coil is fed with an AC voltage, the iron core ensures that all the flux generated in the primary coil also passes through the secondary coil. Figure 6.11b illustrates the symbol used in circuit diagrams for an iron-core transformer.

Ideal transformers are 100% efficient; real transformers are often more than 99% efficient, and for this reason power losses within the transformer can be ignored in calculations. The transformer equation can be written in different ways but is based on:

$$\frac{V_p}{V_s} = \frac{N_s}{N_p}$$

where:

V_p is the primary voltage (V)

V_s is the secondary voltage (V)

N_s is the number of turns in the secondary coil

N_p is the number of turns in the primary coil.

A **step-up transformer** increases the secondary voltage compared with the primary voltage. A **step-down** transformer decreases the secondary voltage compared with the primary voltage.

The transformer equation can also be written in terms of current, i.e.:

$$V_p I_p = V_s I_s$$

Transformers will not work with DC voltage since it has a constant, unchanging current that creates no change in magnetic flux. A transformer works on the basis of a changing current (AC voltage) in the primary coil inducing a changing magnetic flux, which therefore induces a current in the secondary coil. The AC electrical supply from a generator is easily stepped up or down by transformers, so AC is the preferred form of electrical energy in large-scale transmission systems.

Large-scale transmission systems involve current travelling large distances and so even relatively good electrical conductors such as copper have a significant resistance. The efficient transmission of the electrical energy with the least amount of power loss over that distance is therefore important to consider.

> Electric power is the rate, per unit time, at which electrical energy is transferred by an electric current. The power supplied in an electrical circuit is given by $P = VI$.

Electrical power loss is proportional to the square of the current:

$$P = I^2 R$$

where:

P is the power (W)

I is the current (A)

R is the resistance (Ω).

Using this equation, it can be recognised that electrical engineers use very high voltages ($P = VI$) in order to keep the currents small and power losses as heat to a minimum.

Applications of the motor effect

MOTORS

The force produced on a current-carrying conductor leads to very important practical applications, such as an electric meter or the basic principle of operation of electric motors. Recall the magnetic force experienced by a current-carrying conductor, moving at an angle θ to the magnetic field, is given by $F = lIB\sin\theta$. As discussed above, if a current and a magnetic field are both present, the magnetic force acting on each side of a current-carrying square wire coil in a magnetic field will cause it to turn. The turning effect of a magnetic field on a coil of conducting wire is called electromagnetic **torque**. Torque is the turning effect of any force; for example, pushing on a swinging door.

Torque is defined as:

$$\tau = nIA_{\perp}B = nIAB\sin\theta$$

where:

τ is the torque in newton metres (N m)

n is the number of turns

I is the current (A)

$A_{\perp}$ is the area vector perpendicular to the plane of the coil (m^2)

B is the strength of the magnetic field (T)

θ is defined as the angle between the area vector normal to the plane of the coil and the magnetic field.

As with the equation for magnetic flux discussed above, the variable A in the equation for torque actually describes a vector quantity which is directed normal to the plane of the area. This means that when the plane of the coil is parallel to the magnetic field, the area vector is actually directed perpendicular and the angle $\theta = 90°$. Make sure you know which angle you are using when completing your calculations, and remember that drawing a diagram of the situation you are analysing can help you visualise what you need to calculate. Figure 6.12 describes each of the variables required to calculate the torque.

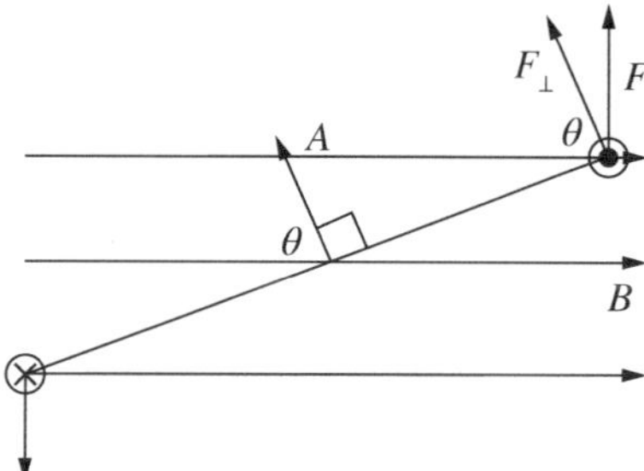

FIGURE 6.12 The angle given in the torque equation for a motor is between the area vector *A* normal to the plane of the coil and the magnetic field, *B*.

When the area vector is perpendicular to the magnetic field (θ is 90°), the full magnitude of the magnetic force is experienced. When the area vector is parallel to the magnetic field, the current-carrying coil experiences no force (θ is 0°).

A DC motor is a device that makes use of the motor principle but contains a switch assembly on the rotating coil shaft that allows the direction of the current through the coil to be reversed every 180°. This switch is called a commutator. It allows the DC motor to keep rotating because the direction of current, and hence torque, is reversed each half turn by the commutator. This is shown in Figure 6.13.

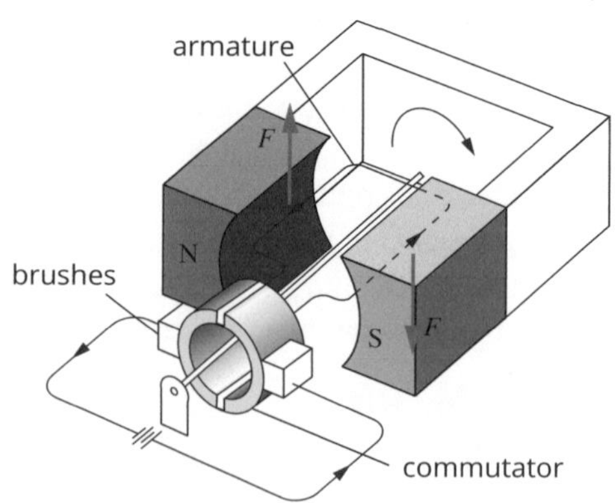

FIGURE 6.13 The main parts of a simple but practical single-coil DC electric motor

The coils are wound around a soft iron core to increase the magnetic field that passes through them. The whole arrangement of core and coils is called an **armature**. The armature of a practical motor consists of many coils that are fed current by the commutator when they are in the position of maximum torque. The total torque will be the sum of the torques on all the individual coils.

Another type of motor is the AC induction motor. This motor rotates because of the interaction of the magnetic fields of the rotor and the stator. In a DC motor, the magnetic field is stationary (the stator), and the coil that carries the electric current rotates (the rotor). The stator used in an AC induction motor is made of electromagnetic coils that can produce greater magnetic field strengths than a permanent magnet. The stator creates a rotating magnetic field that induces a current in the conductors of the rotor. Through application of Faraday's and Lenz's laws, the rotating magnetic field of the stator effectively pulls the rotor.

GENERATORS

A machine that converts mechanical energy into electrical energy is called a **generator** or **alternator**. The electric generator is probably the most important practical application of Faraday's discovery of electromagnetic induction. This machine relies on the induction principle between a coil and a magnetic field. The principle of electric power generators is the same whether the result is alternating current or direct current. Relative motion between a coil and a magnetic field induces an emf in the coil.

The construction of a generator or an alternator is very similar to that of an electric motor. A coil is rotated in a magnetic field which will produce an alternating induced current in the coil. How that current is harnessed will determine if the device is an AC alternator or a DC generator.

 ISBN 978 1 4886 1936 6

An AC alternator has slip rings that transfer the alternating nature of the current in the coil to the output (shown in Figure 6.14a). A DC generator has a split ring commutator to reverse the current direction every half turn so that the output current is always in the same direction (shown in Figure 6.14b).

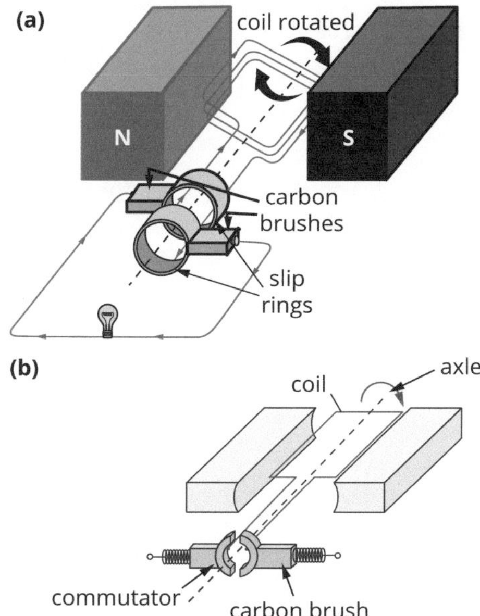

FIGURE 6.14 (a) AC alternator and (b) DC generator

An AC alternator produces an alternating current that varies sinusoidally over time with the change in magnetic flux (the sinusoidal shape is shown in Figure 6.15). The alternating current produced by power stations and supplied to cities varies sinusoidally at a frequency of 50 Hz. The peak value of the voltage of domestic power (V_p) is ±340 V, and the peak-to-peak voltage (V_{p-p}) is 680 V. This is shown in Figure 6.15.

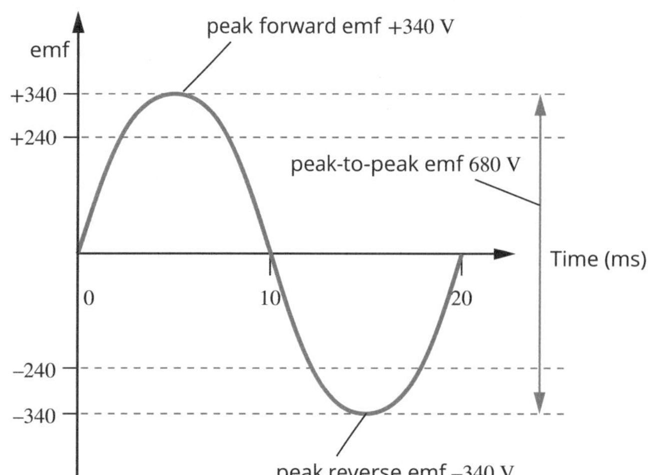

FIGURE 6.15 The voltage in Australian power points oscillates between +340 V and –340 V, 50 times each second. The value of a DC supply that would supply the same average power is 240 V.

Sometimes it is more useful to know the average power produced in a circuit. The average power can be obtained by using the **root mean square** or rms value. The rms value of domestic mains voltage in Australia is 240 V. The root mean square voltage, V_{rms}, is the value of an equivalent steady voltage (DC) supply that would provide the same average power as the AC supply.

$$V_{rms} = \frac{V_p}{\sqrt{2}}$$

where V_p is the peak voltage (V).

Once V_{rms} is known, this can be used to find the average power in a resistive AC circuit. The formula is:

$$P = V_{rms} I_{rms}$$
$$= \frac{1}{2} \times V_p \times I_p$$

WORKSHEET 6.1

Knowledge review—checking up on charges

1 Two positive charges, *P* and *Q*, are placed a distance *r* apart. Charge *Q* experiences a force of *F*.

a In terms of *r*, how far from *P* would *Q* need to be placed to experience a force of 4*F*?

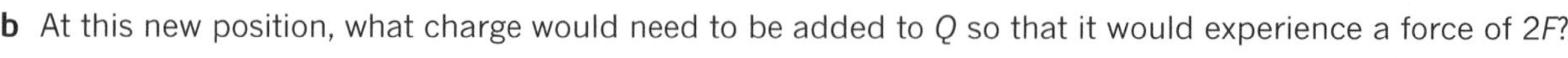

b At this new position, what charge would need to be added to *Q* so that it would experience a force of 2*F*?

2 A 24 V truck battery is connected to two resistors in series as shown.

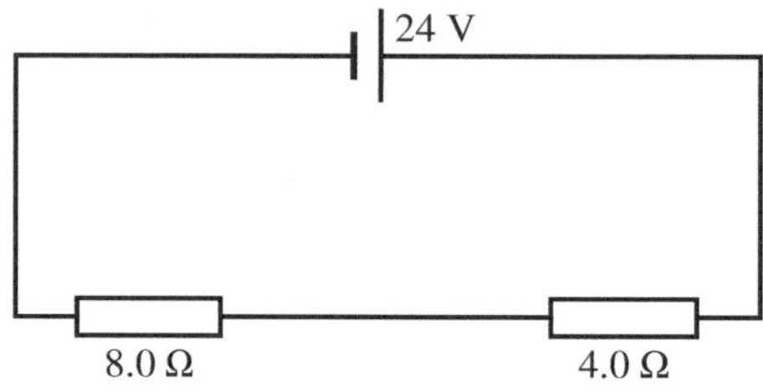

a What is the total resistance in the circuit?

b What is the current flowing through the 8.0 Ω resistor?

c What is the potential difference across the 4.0 Ω resistor?

d What current flows through the battery if another 8.0 Ω resistor is placed in parallel with the 4.0 Ω resistor?

3 A wire has a current vertically upwards through a hole (i.e. out of the page in the diagram below) in a horizontal platform in a region where the Earth's magnetic field is horizontal and has a strength of 5.0×10^{-5} T.

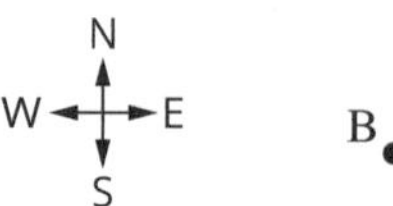

D

ISBN 978 1 4886 1936 6

a Points A to D are each 12.0 cm from the wire. A compass is placed in turn at each of the points. At one point the compass points directly north, at another exactly north-east and at another the compass keeps moving around without settling down in any one direction. Identify each of these three points.

Points directly north = ____________ Points north-east = ____________ Compass keeps moving = ____________

b What is the size of the net field at each of these three points?

c Find the size of the current in the wire.

4 A 12.0 V DC power supply is attached to a 56.0 Ω resistor.

a What current does the power supply carry?

b How much power is being expended in the circuit by the power supply?

c How do these quantities change if a 33.0 Ω resistor is now connected in series with the first resistor?

5 A 14.0 cm long coil around a cardboard tube has 600 turns and carries a current of 1.20 A.

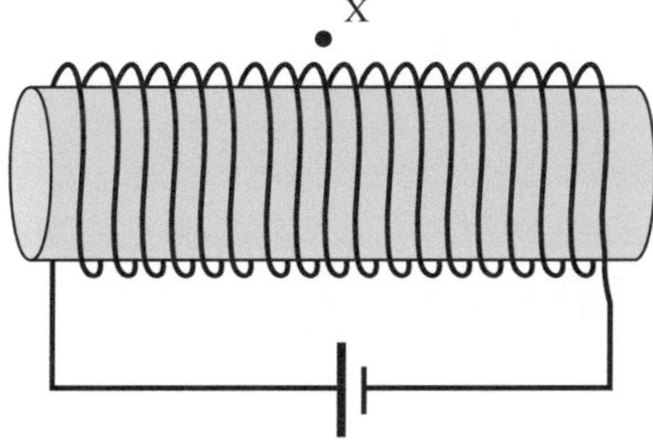

a What is the direction of the magnetic field in the middle of the coil?

b What is the magnetic field strength at the mid-point inside the coil?

c What is the field strength at point X, just outside the middle of the coil?

6 The diagram below depicts four equipotential lines (lines of equal potential) in an electric field.

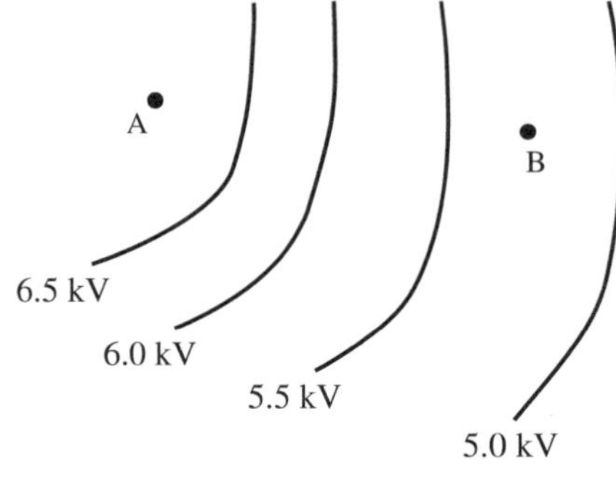

a On the diagram draw five electric field lines, indicating their direction.

b At which point, A or B, is the field strongest?

c How much work would be done by the field moving a charge of 3.4×10^{-17} C between the left and the right equipotential lines?

7 At a certain point in a circuit, 4.5×10^{17} electrons per second are recorded moving from left to right through a 25 cm length of wire.

a What is the direction of the conventional current?

b What is its size?

c If every electron has 8.0×10^{-18} J of work done on it, find the potential difference across the length of wire.

d What is the size of the electric field in the wire?

 ISBN 978 1 4886 1936 6

WORKSHEET 6.2

Deflecting electrons

A group of students set up an old cathode ray oscilloscope tube to study the behaviour of electrons in a magnetic field. They placed a 'horseshoe' electromagnet around the straight part of the tube as shown. A DC power supply produces a current, I, in the electromagnet.

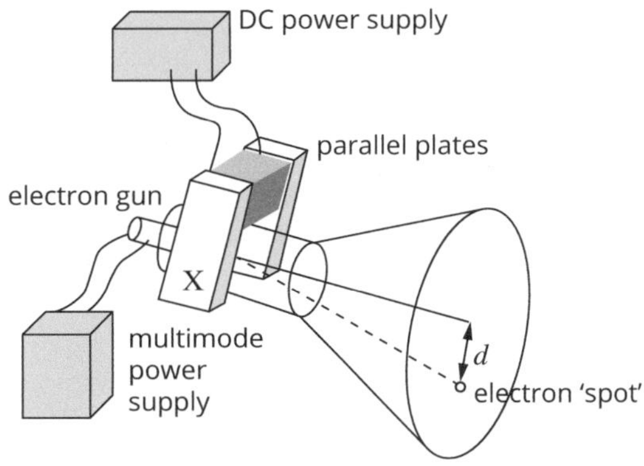

Electrons are produced in the electron gun by means of thermionic emission. The electrons are accelerated by means of a high DC voltage applied across a pair of plates just outside the gun. Then the focusing electrodes within the cathode ray tube produce a beam which is focused at one point on the screen, creating a bright spot. In the absence of any external magnetic field the spot is in the centre of the screen. When the field acts on the electron beam, the spot is deviated by a distance, d, from the centre.

a Research and briefly explain how thermionic emission works.

b Does X represent the north or south pole of the electromagnet? Explain your answer.

During the investigation, the current, I, in the electromagnet and the accelerating potential, V, were both varied systematically. At each stage, the deflection distance, d, was recorded. The students realised that the electrons followed a circular arc for the time they were between the poles of the magnet. By careful scale drawing and use of geometry they were able to show that the radius of curvature was given by the relationship $r \approx \frac{85}{d}$.

The tables below display their results.

TABLE 1 Variation of current with a constant accelerating potential, V, of 4.0 kV

I (A)	d (cm)
0.21	0.9
0.34	1.7
0.43	2.1
0.57	2.8
0.70	3.1
0.84	4.1

TABLE 2 Variation of voltage with a constant electromagnet current, I, of 0.50 A

V (kV)	d (cm)
1.8	2.5
2.1	2.3
2.9	2.0
3.7	1.8
4.0	1.7
5.5	1.5

The students were keen to test whether these results were consistent with the equations they had studied for charged particles travelling in magnetic fields. First, they had to re-arrange the equations.

c Identify the independent and dependent variables in this investigation.

Important relationships:

- The kinetic energy gained by each electron in the gun is given by the equation $K = qV = \frac{1}{2}mv^2$.
- The radius of curvature is given by the relationship, $F_c = \frac{mv^2}{r}$.
- The magnetic force acting on each charged particle is given by $F = qvB$.
- The radius of curvature is given by $r \simeq \frac{85}{d}$.
- The strength of the magnetic field will be proportional to the current: $B \propto I$.

d By manipulating these equations, show that $d \propto I$. (Hint: Find the relationship between B and r by equating the magnetic force on a charge to the centripetal force required. Then replace B with constant $\times I$ and r with $\frac{85}{d}$.)

e Plot a graph of d vs I. Does it support the relationship $d \propto I$?

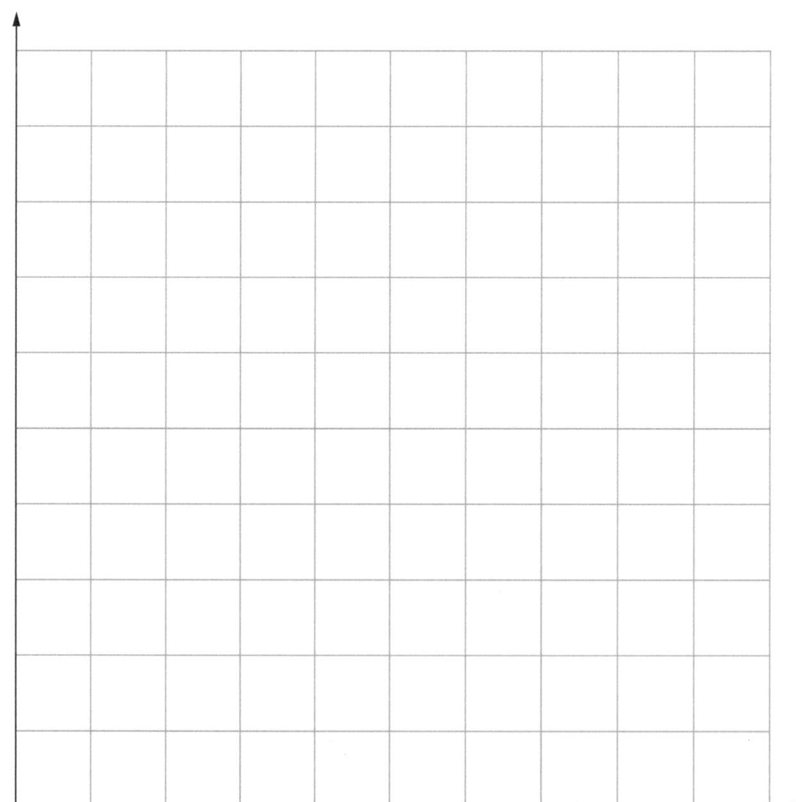

 ISBN 978 1 4886 1936 6

f Demonstrate that $r \propto \sqrt{V}$. (Hint: First find a relationship between r and v by equating the magnetic force on a charge to the centripetal force required. Then find one between v and V. Use these two relationships to find the one between r and V.)

g To see if this relationship is demonstrated, add extra columns for values of r and $\sqrt{V}$ to Table 2 above. Sketch r vs $\sqrt{V}$. Does it support the relationship $r \propto \sqrt{V}$?

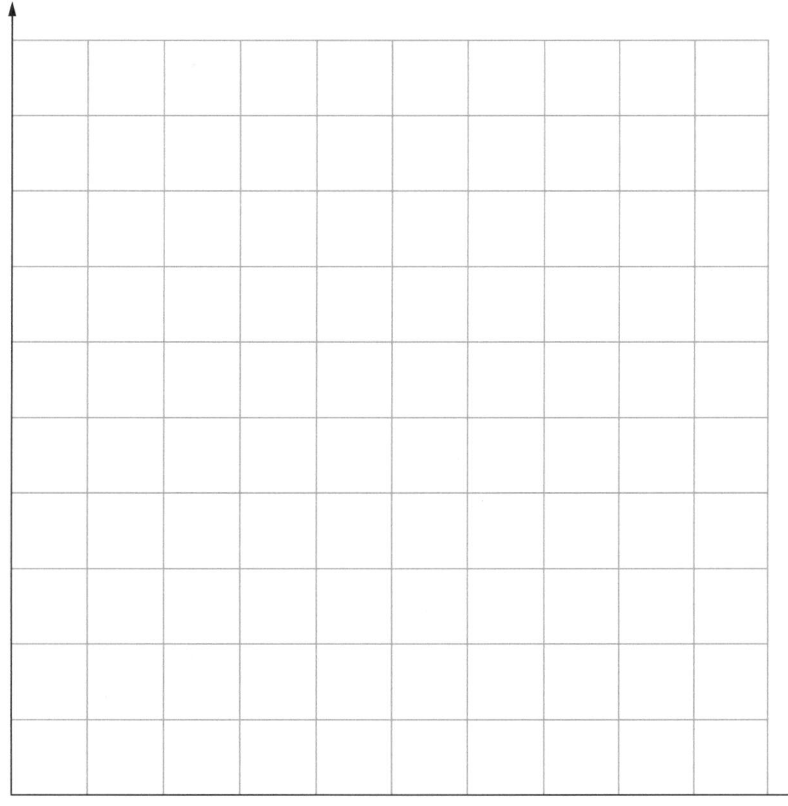

RATING MY LEARNING	My understanding improved	Not confident ◄—► Very confident ○ ○ ○ ○ ○	I answered questions without help	Not confident ◄—► Very confident ○ ○ ○ ○ ○	I corrected my errors without help	Not confident ◄—► Very confident ○ ○ ○ ○ ○

WORKSHEET 6.3

Force and current

Ethan conducted an experiment to evaluate the magnetic field strength of a coil. This coil had dimensions of 40 cm high and 5 cm wide, and had 200 turns of wire. The experiment involved connecting the coil of wire to a battery and an ammeter, which is suspended from a spring balance between the poles of two magnets. The spring balance readings were recorded for different currents. The apparatus and results are shown below:

Force (N)	Current (A)
2.9	0.5
3.3	1
3.7	1.5
4.1	2
4.5	2.5
4.9	3
5.3	3.5
5.7	4
6.1	4.5
6.5	5

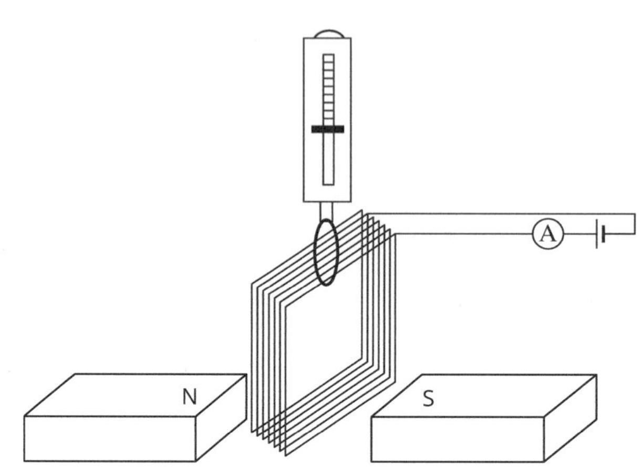

1 State the mathematical relationship of a force acting on any conductor when placed in a magnetic field. Identify the known variables from the question.

2 **a** Graph the data above in the space provided. According to these results, what will the force be when the current is zero?

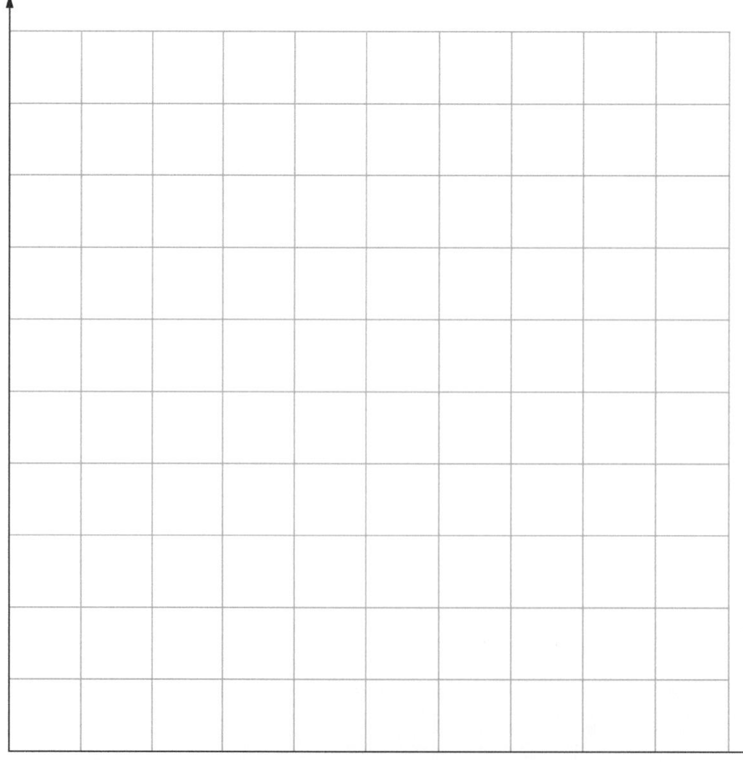

 ISBN 978 1 4886 1936 6

b Identify what this additional force is.

c Determine the mass of the coil.

d Using your results from question 2c, evaluate the magnitude of the magnetic field strength.

RATING MY LEARNING	My understanding improved	Not confident ⟷ Very confident ○ ○ ○ ○ ○	I answered questions without help	Not confident ⟷ Very confident ○ ○ ○ ○ ○	I corrected my errors without help	Not confident ⟷ Very confident ○ ○ ○ ○ ○

WORKSHEET 6.4

The motor effect

The diagrams below depict a simplified view of a single square loop, SPQR, in a vertical, uniform magnetic field. The loop is free to rotate about an axis inside the coil. The diagram on the right is a side-on view along the axis. Each side of the loop is 4.0 cm in length and the strength of the magnetic field is 8.0 T. The loop carries a current of 200 mA and is shown at an angle of 35° to the horizontal.

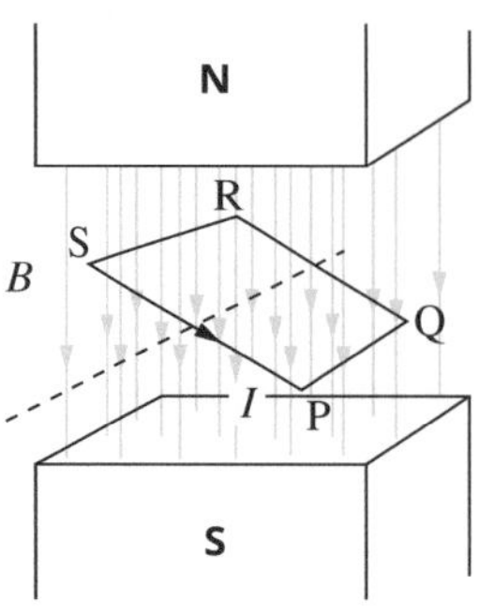

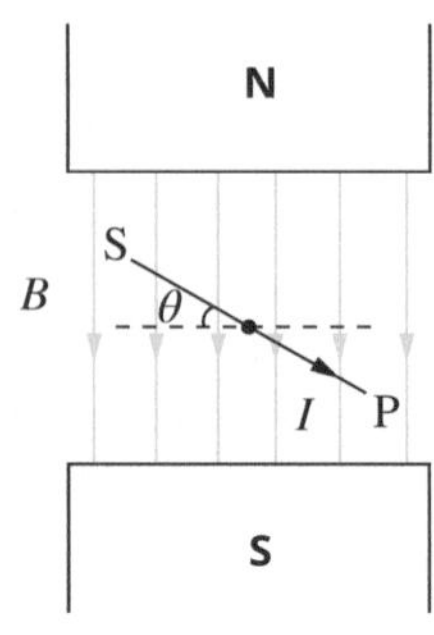

1 What is the magnetic force acting on side PQ? (Include a direction in your answer.)

2 What is the magnetic force acting on side RS? (Include a direction in your answer.)

3 What is the magnitude of the torque on the coil?

4 What effect will this torque have on the loop?

5 Calculate the magnetic forces acting on sides SP and QR. (Include the direction in your answer.)

6 What effect will these forces have on the loop?

7 At what angle will the torque acting on the loop be a maximum?

ISBN 978 1 4886 1936 6

8 What happens to the forces on sides PQ and RS (and to the torque) when the angle θ is 180°?

9 Describe what happens to the forces on PQ and RS (and to the torque) when the angle θ exceeds 180°.

10 If this loop is to continue to be rotated by the magnetic forces, what needs to happen to the current? Describe how this action can be applied to the loop. What mechanism can you use to achieve this?

11 Explain three further modifications or additions that could be made to this arrangement to turn it into a practical electric motor.

RATING MY LEARNING	My understanding improved	Not confident ◄──► Very confident ○ ○ ○ ○ ○	I answered questions without help	Not confident ◄──► Very confident ○ ○ ○ ○ ○	I corrected my errors without help	Not confident ◄──► Very confident ○ ○ ○ ○ ○

WORKSHEET 6.5

Small-scale power transmission

A farmer has just bought himself a wind turbine which produces a maximum power of 12 kW at 250 V rms. When the wind is not blowing well enough for this output, he plans to buy power from the grid. He also plans to sell power to the grid when he can't use all that is generated.

He placed the turbine on the top of a hill behind his farmhouse, at a distance of 326 m. He buys some twin-core power cable from the local hardware store and strings it up on poles as shown below.

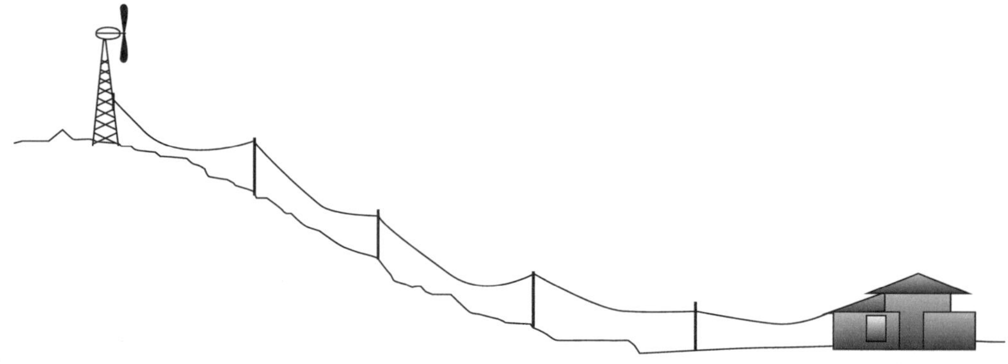

1 Each of the two cores within the cable has a resistance of $0.004\,\Omega\,m^{-1}$. What is the total resistance of the transmission circuit between the turbine and the house?

2 How much current does the turbine produce under optimum conditions (i.e. when operating at maximum power)?

3 Calculate the power loss in the cables.

4 What is the voltage drop across the transmission cables?

5 What voltage is available at the house?

After he discovers these issues, the farmer rings up his old Physics teacher and asks her for help. She, in turn, advises that step-up and step-down transformers with turn ratios of 25:1 will need to be installed so that the power can be transmitted at a higher voltage with less loss.

6 Where should these be installed and why?

 ISBN 978 1 4886 1936 6

WORKSHEET 6.5

A further suggestion was that another 326 m of the same twin-core cable can be strung up alongside the original cable. This way the two cores in each cable can be used in parallel so that one twin cable provides the outwards path for the current and the other provides the return path. For the remaining questions, assume that the turbine is operating at maximum power.

7 What will be the new resistance of the transmission line?

8 What will the voltages be at the input and output of the first transformer?

9 How much current will the line be carrying?

10 How much power will now be lost in transmission?

11 What will be the voltage drop across the transmission lines now?

12 What voltage will be available at the input of the second transformer?

13 What voltage will now be available at the house?

RATING MY LEARNING	My understanding improved	Not confident ◄——► Very confident ○ ○ ○ ○ ○	I answered questions without help	Not confident ◄——► Very confident ○ ○ ○ ○ ○	I corrected my errors without help	Not confident ◄——► Very confident ○ ○ ○ ○ ○

WORKSHEET 6.6

Back to motors and generators

1 The figure below shows the main features of a DC motor. Label each of the four features indicated and describe their purpose.

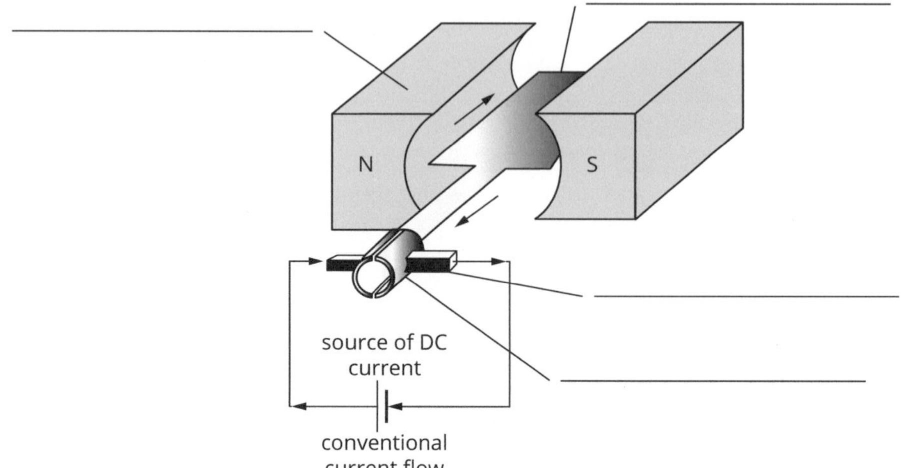

2 The DC motor shown has the magnetic poles shaped to create a radial field. What advantage does this have over flat magnetic poles?

3 A simple motor can be a generator and a simple generator can be a motor. Describe, in point form, the similarities and differences between simple motors and generators.

Similarities	Differences

4 An electric motor rotates due to the torque applied on the coil by the current through the coil interacting with the magnetic field of the motor. However, the action of the coil moving through the motor's magnetic field also causes a voltage to be induced. What is the name for this induced voltage and in what direction will it act?

5 Around 95% of the world's electric motors run on AC rather than DC current. In a DC motor, the magnetic field is stationary (the stator), and the coil that carries the electric current rotates (the rotor). An AC induction motor works by producing a rotating magnetic field. Describe the structure and function of the rotor and stator in an AC induction motor.

 ISBN 978 1 4886 1936 6

6 Describe one advantage and one disadvantage of an AC induction motor.

7 The following figure shows a simple DC generator and an AC alternator.

(a)

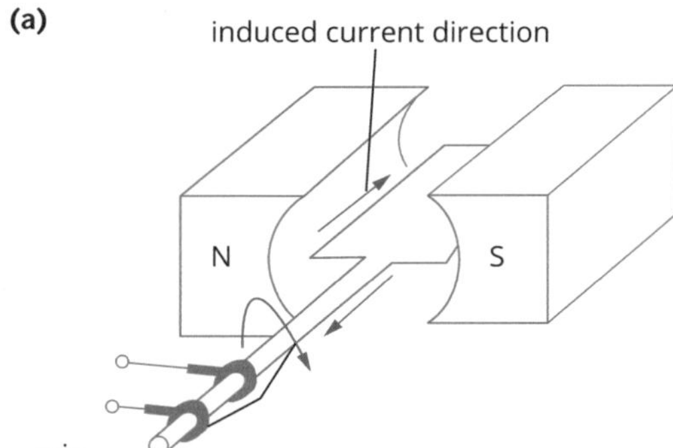

(b)

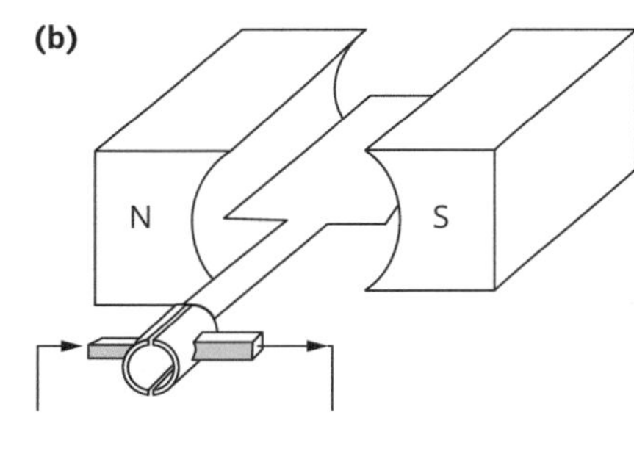

Which of these generators is the AC alternator and which is the DC generator? Explain your reasoning.

8 Sketch the output of the induced voltage from each of the generators in the space provided.

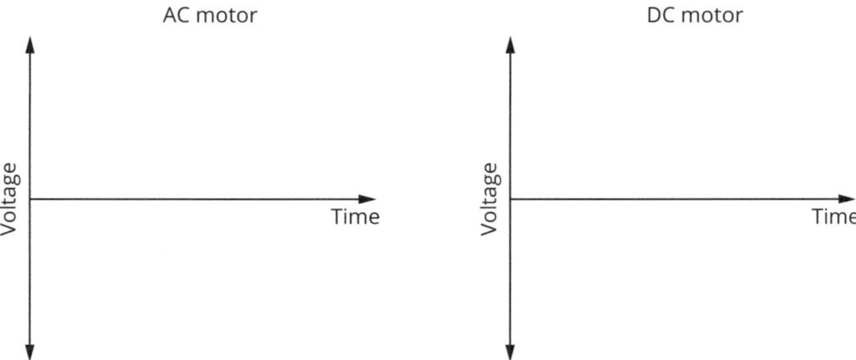

9 Magnetic brakes can produce a strong braking force without the need for physical contact required in standard drum or disc brakes. In the space provided, describe with the use of diagrams the operation of a magnetic brake.

RATING MY LEARNING	My understanding improved	Not confident ⟷ Very confident ○ ○ ○ ○ ○	I answered questions without help	Not confident ⟷ Very confident ○ ○ ○ ○ ○	I corrected my errors without help	Not confident ⟷ Very confident ○ ○ ○ ○ ○

WORKSHEET 6.7

Literacy review—understanding electromagnetism

1 Define each of the following terms.

Term	Definition
commutator	
magnetic flux	
electromagnetic induction	
electromotive force	
solenoid	
transformer	
back emf	
torque	

2 Fill in the blanks from the following list of terms. Some terms may be used more than once.

coil	displacement	conductors	magnetic flux	motor	magnetic
attract	Faraday's	field	repel	right-hand	flux

a ______________ law defines the relationship between the number of turns in a coil, N, and the rate of change in ______________ ______________, Φ. ______________ ______________ is related to the strength of the ______________ field, the area enclosed by the wire loop, and the angle between them. We can say that the ______________ is proportional to the strength of the magnetic field.

b If two parallel ______________ carry current in the same direction, the forces ______________. If two parallel ______________ carry current in the opposite direction, the forces ______________. This can be proven using the______________ rule.

RATING MY LEARNING	My understanding improved	Not confident ◀——▶ Very confident ○ ○ ○ ○ ○	I answered questions without help	Not confident ◀——▶ Very confident ○ ○ ○ ○ ○	I corrected my errors without help	Not confident ◀——▶ Very confident ○ ○ ○ ○ ○

ISBN 978 1 4886 1936 6

WORKSHEET 6.8

Thinking about my learning

On completion of Module 6: Electromagnetism, you should be able to describe, explain and apply the relevant scientific ideas. You should also be able to interpret, analyse and evaluate data.

1 The table lists the key knowledge covered in this module. Read each and reflect on how well you understand each concept. Rate your learning by shading the circle that corresponds to your level of understanding for each concept. It may be helpful to use colour as a visual representation. For example:

- green—very confident
- orange—in the middle
- red—starting to develop.

Concept focus	**Rate my learning**				
	Starting to develop ◄				► Very confident
The motion of charged particles in electric fields	○	○	○	○	○
The motion of charged particles in magnetic fields	○	○	○	○	○
The force on a current-carrying conductor in a magnetic field	○	○	○	○	○
Magnetic flux and induced emf	○	○	○	○	○
Faraday's law	○	○	○	○	○
Lenz's law	○	○	○	○	○
Step-up and step-down transformers	○	○	○	○	○
Eddy currents	○	○	○	○	○
Back emf	○	○	○	○	○
Motors	○	○	○	○	○
Generators	○	○	○	○	○

2 Consider points you have shaded from starting to develop to middle-level understanding. List specific ideas you can identify that were challenging.

3 Write down two different strategies that you will apply to help further your understanding of these ideas.

PRACTICAL ACTIVITY 6.1

A current-carrying conductor in a uniform magnetic field—force versus current

Suggested duration: 30 minutes

INTRODUCTION

A current-carrying wire in a magnetic field experiences a force that is usually referred to as a magnetic force. The magnitude and direction of this force depends on four variables:

- the magnitude of the current (I);
- the length of the wire (l);
- the strength of the magnetic field (B); and
- the angle between the field and the wire (θ).

This magnetic force can be described mathematically by the equation:

$$F = BIl \sin\theta.$$

PURPOSE

To investigate the relationship between force and current for a current-carrying conductor in a magnetic field.

PROCEDURE

1 Set up the current balance following the manufacturer's instructions.

2 With no current flowing, either use small masses to balance the beam or record the mass using a beam or electronic balance. Record the mass in the Data and analysis section. If you are using an electronic balance, tare the balance to read zero with no current flowing.

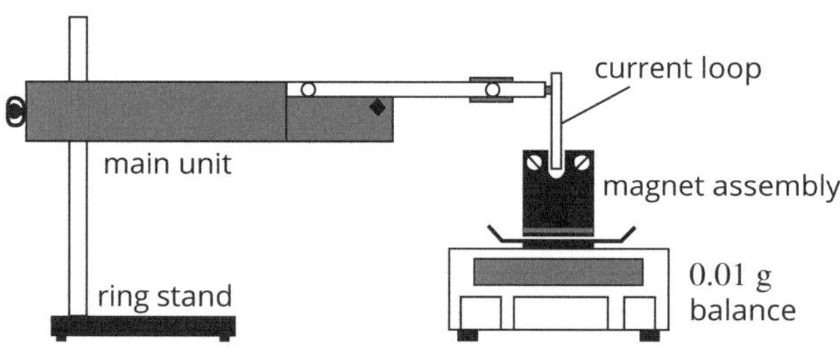

3 Position the magnet assembly around the end of the straight wire at the end of the current balance. This may be a magnet or an air-core solenoid. If using a solenoid, connect the second power supply to the solenoid and set it at approximately 5 V.

4 Rebalance the current balance by adding or removing small masses from the opposite end of the current balance, or record the new mass if using a beam balance. If using an electronic balance, tare the balance back to zero.

5 Connect the power supply and ammeter to the current balance. If using a solenoid to provide the magnetic field, a second power supply should be connected to the solenoid to allow separate control.

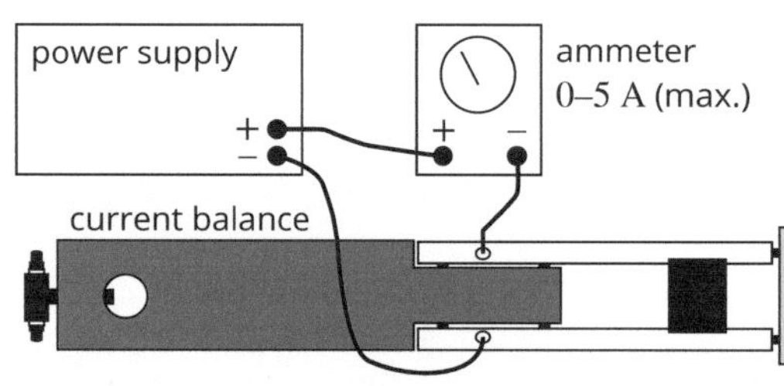

MATERIALS

- current balance*
- DC power supply supplying up to 5 A
- DC ammeter, current sensor or current probe measuring up to 5 A
- 0.01 g resolution beam, electronic balance or small milligram masses
- magnets or air-cored solenoid and 0–20 V power supply
- retort stand or lab stand
- connecting wires with banana plug connections

*Current balances of different design are available from various suppliers. Follow the manufacturer's specific set-up and safety instructions when conducting this activity. Effective balances can also be student-made using copper wire and stiff card or plastic sheets (non-conductive).

As the coil and beam conductors carry current, they will become warmer and their resistance will rise and the current will fall over time. If available, use 'constant current' power sources.

To prevent burning of the balance contact points and possible damage to the circuit board, do not pass currents in excess of the manufacturer's recommendation through the current balance.

To avoid permanent damage to the current-measuring device, be careful to keep the supplied current below the maximum limit of the measuring device.

 ISBN 978 1 4886 1936 6

6 Check the maximum current that your current sensor or ammeter and current balance can safely use. Set the current from the power supply to give a reading of 1/10 of the maximum range. For example, if the maximum range is 1 A, adjust the current from the power supply to read 0.1 A on the current sensor or ammeter. For a maximum range of 5 A, adjust the current to show a reading of 0.5 A on the current sensor or ammeter.

7 Rebalance the current balance or read the new mass on the balance. Record this value in the table in the Data and analysis section.

8 Increase the current in 1/10 increments (i.e. 0.1 A for a 1 A range). Balance the current balance with small weights or record the mass on the balance for each current setting. Do not exceed the maximum range of the ammeter or current sensor, or the safe working range of the current balance.

DATA AND ANALYSIS

1 When investigating the relationship between force and current, what angle should the wire along the end of the current balance make with the magnetic field? Why?

2 Why can the effect of the connecting wires along the length of the current balance be ignored?

3 Calculate the force corresponding to each measurement of mass and record it in the corresponding column of Table 1. How is the force calculated?

TABLE 1 Force versus current

Current (A)	Mass (g)	Force (N)
0		

4 Plot a graph of force versus current in the space provided, using a suitable scale.

Graph of force (*F*) versus current (*I*)

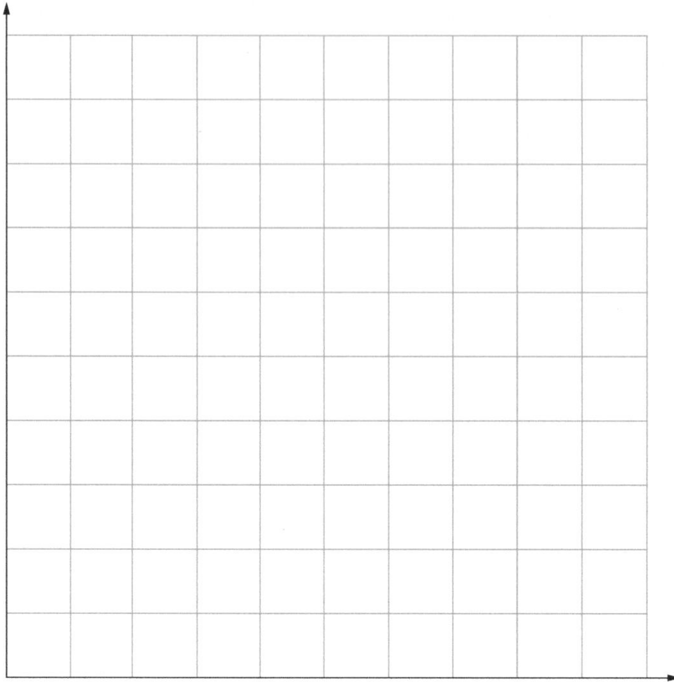

5 Determine the equation of the relationship between the force and current for your data.

CONCLUSION

1 What is the nature of the relationship between these two variables?

2 What does this tell you about how changes in the current will affect the force acting on a wire that is inside a magnetic field?

3 What is the significance of the gradient of the graph? How reliable was the value you found?

FURTHER INVESTIGATION

You can investigate the relationship between the force and the strength of the magnetic field by repeating this activity, but instead of varying the current, keep the current constant and vary the strength of the magnetic field. Do this by either adding more magnets, or, if you are using an air-cored solenoid, by varying the voltage to the solenoid. Variations in the voltage to the solenoid should be made in 1 V increments.

You can investigate the relationship between the force and the length of wire in the magnetic field by again repeating this activity, but instead of varying the current, keep the current constant and vary the length of the conductor in the magnetic field.

RATING MY LEARNING	My understanding improved	Not confident ◄——► Very confident ○ ○ ○ ○ ○	I answered questions without help	Not confident ◄——► Very confident ○ ○ ○ ○ ○	I corrected my errors without help	Not confident ◄——► Very confident ○ ○ ○ ○ ○

 ISBN 978 1 4886 1936 6

PRACTICAL ACTIVITY 6.2

Parallel wires—quantitative data analysis

Suggested duration: 40 minutes

INTRODUCTION

Any current-carrying conductor will have a magnetic field associated with it. This means that between any two parallel current-carrying conductors the magnetic field of each will interact. This is something that particularly needs to be considered in designing transmission lines that will be in close proximity.

PURPOSE

To quantitatively analyse the way two parallel current-carrying conductors interact.

PROCEDURE

In a typical installation for a high-voltage transmission installation of 22 kV, a bundled cable of aluminium alloys around a galvanised steel core is used. The mass per unit length of the cable is 1540 kg km^{-1}. The maximum continuous current the cable can supply in still air is 110 A.

Two 1 m sections are placed on an adjustable rack in parallel, one piece vertically above the other. The wires are connected in series. The arrangement is shown in the figure below.

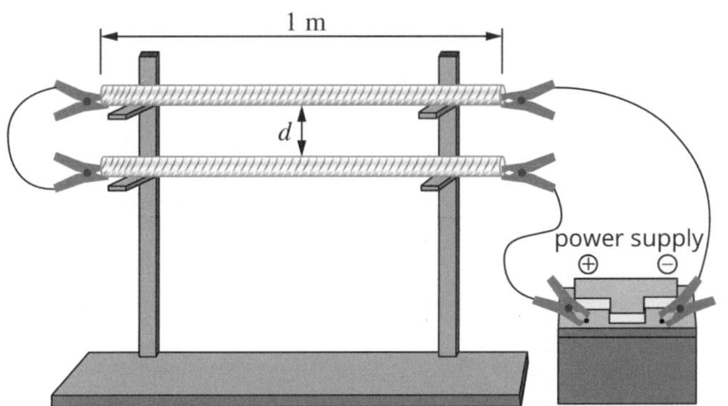

Complete an analysis of the effects felt by the interaction of the two wires.

DATA AND ANALYSIS

1 What will be the ratio of the current in the top wire to that in the bottom wire?

2 Will the wires attract or repel? Explain your answer.

3 The cables are now brought toward each other and the force per unit length between them is measured. Assuming that the power supply is supplying a constant 10 A of current, complete Table 1 for the variation in distance.

TABLE 1 Force per unit length with variation in distance

Distance *d* (cm)	$\frac{F_B}{l}$ (N m^{-1})
30	
25	
20	
15	
10	
5	
1	

PRACTICAL ACTIVITY 6.2

4 Plot a graph of F_B versus d from the data in Table 1. What relationship does the shape of the graph represent?

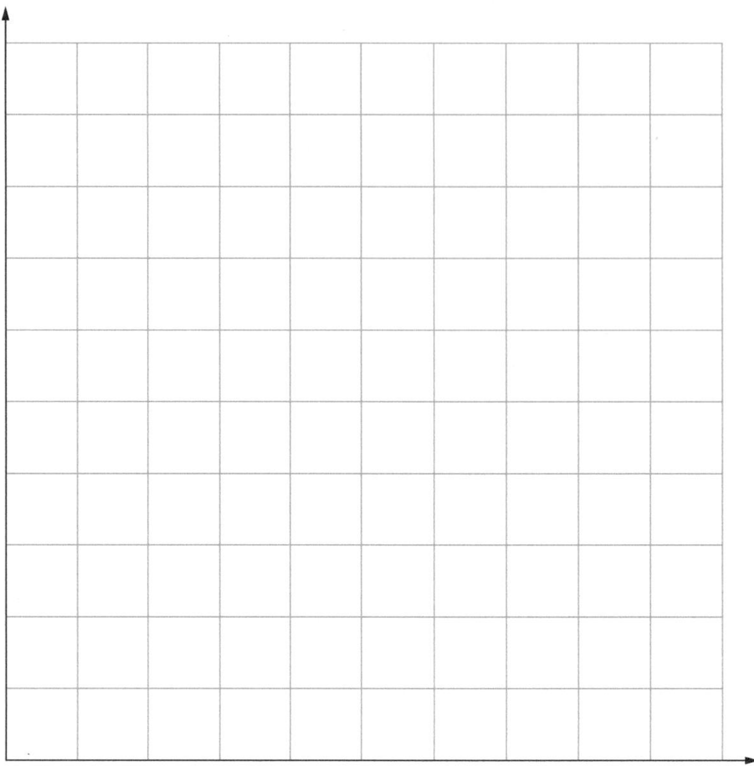

5 At what distance apart would the force due to the magnetic field around each wire be sufficient to balance the mass of the top wire? Assume that the wires are straight and the lower wire is unable to move away.

6 The cable is designed to support a maximum continuous current of 110 A. If the cable were carrying this current, at what distance would the force between the wires balance the mass of the top wire?

7 If the force experienced by the top wire is exactly balancing the force due to gravity, would the force experienced by the bottom wire be larger, the same or smaller? Explain your answer.

RATING MY LEARNING	My understanding improved	Not confident ◄——► Very confident ○ ○ ○ ○ ○	I answered questions without help	Not confident ◄——► Very confident ○ ○ ○ ○ ○	I corrected my errors without help	Not confident ◄——► Very confident ○ ○ ○ ○ ○

ISBN 978 1 4886 1936 6

PRACTICAL ACTIVITY 6.3

Electromagnetic induction—the direction of the induced current in a wire

Suggested duration: 20 minutes

INTRODUCTION

Just three years after Faraday's discovery of electromagnetic induction, German physicist Heinrich Lenz discovered a simple principle by which the direction of the induced emf could be found. Lenz's law states that a current that is induced in a loop will create a flux which will oppose the change in flux that created the current. The right-hand rule can be used to visualise the direction of the induced current. In this activity the direction of the current induced in a wire by a magnetic field is investigated.

MATERIALS

- galvanometer or current sensor
- horseshoe magnet
- electrical leads with alligator clips

PURPOSE

To investigate the direction of the current induced in a wire by a magnetic field.

PROCEDURE

1 Connect a long copper wire to a galvanometer or current sensor using two leads with alligator clips. If using a current sensor, follow the directions of the manufacturer to set up an analogue display of current. Set a sample rate of 20–50 Hz.

2 Hold a part of the wire horizontally between the two poles of a horseshoe magnet as shown in the diagram.

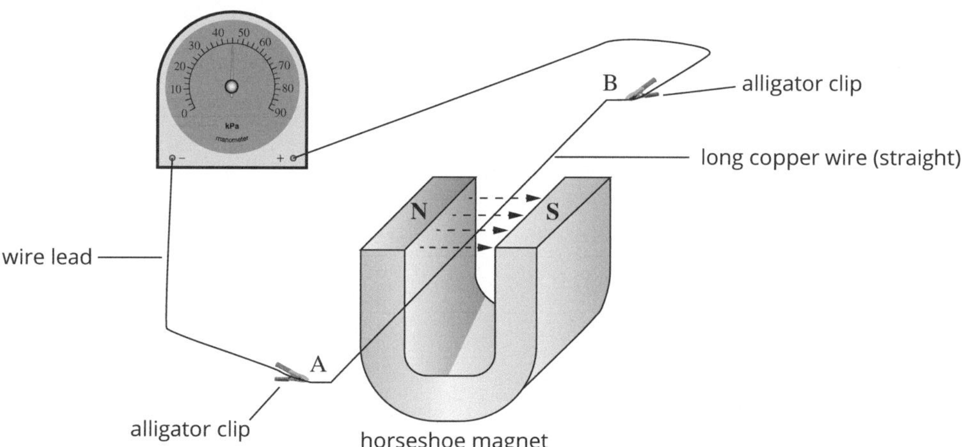

3 Keeping the wire horizontal and parallel to the line A–B, move it vertically up and down within the magnetic field of the magnet. Observe and record the direction of the current induced in the wire as it is moved in each direction. Note the direction in the table in the Data and analysis section.

4 Keeping the wire horizontal and parallel to the line A–B, move it horizontally from one pole of the magnet to the other. Observe and record the direction of the current induced in the wire as it is moved in each direction.

5 Finally, move the wire backwards and forwards along the line A–B, keeping it horizontal. Once again, observe and record the direction of the current induced in the wire as it is moved in each direction.

PRACTICAL ACTIVITY 6.3

DATA AND ANALYSIS

Test	Current direction or no current detected
moving upwards within the magnetic field	
moving downwards within the magnetic field	
moving horizontally to the right	
moving horizontally to the left	
moving away parallel to the direction of the wire	
moving forwards parallel to the direction of the wire	

CONCLUSION

Comment on the following with reference to the figure above and your observations.

1 In which direction does the current flow when the wire is moved vertically upwards?

2 In which direction does current flow when the wire is moved vertically downwards?

3 What current is observed when the wire is moved horizontally within the magnetic field?

4 What current is observed when the wire is moved backwards and forwards along the line A–B within the magnetic field?

5 The relationship between the direction of the magnetic field, the direction of motion of the wire and the direction of the induced current can be shown by three mutually perpendicular arrows.

Write down what each of the arrows in the following figure represents.

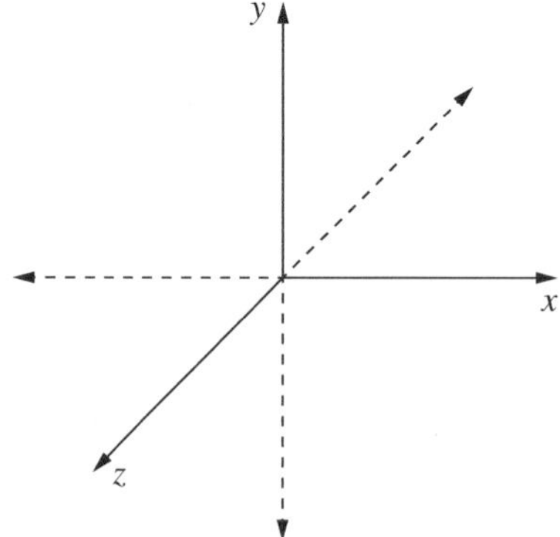

x ______

y ______

z ______

RATING MY LEARNING	My understanding improved	Not confident ◄──► Very confident ○ ○ ○ ○ ○	I answered questions without help	Not confident ◄──► Very confident ○ ○ ○ ○ ○	I corrected my errors without help	Not confident ◄──► Very confident ○ ○ ○ ○ ○

ISBN 978 1 4886 1936 6

PRACTICAL ACTIVITY 6.4

Faraday's law of electromagnetic induction

Suggested duration: 50 minutes

INTRODUCTION

Michael Faraday (1791–1867) discovered a relationship between a changing magnetic flux, Φ, and the potential within a conductor, ε. Known as Faraday's law, this relationship is defined by two key elements: the number of turns in a coil, N, and the change in magnetic flux, Φ. In this activity, Faraday's law will be investigated by using magnets of different strength and increasing the rate at which the magnet passes through the coil.

PURPOSE

To investigate quantitatively the relationship between induced voltage and magnetic flux, time and the number of turns in a coil of wire.

Safety warning

Be careful with magnets. Strong magnets can disrupt electronic devices and severely pinch any skin that comes between them.

Keep magnets away from computer hard drives, USB drives and phones.

MATERIALS

- data-collection system
- voltage sensor (if not available, a CRO or galvanometer can be used for a qualitative investigation)
- 200-, 400- and 800-turn coils (or other coils of known number of turns as available)
- neodymium magnets
- plastic tubes with lids (empty blood-test tubes from a pathology lab are ideal)
- retort stand and clamp
- paper and tape
- no-bounce or foam pad (optional)

PROCEDURE

Neodymium magnets can severely pinch skin and are very fragile. Prior to starting the experiment, place the magnets in the plastic tubes—one magnet in one tube, two in the next and so on. Seal the tubes to prevent the magnets falling out.

Part A: Varying the number of turns

1. Connect the voltage sensor to the data analysis system following the manufacturer's guidelines and start a new experiment in the relevant software. Set the sample rate to at least 200 Hz and display a graph of voltage versus time.
2. Mount the 200-turn coil to the retort stand using the clamp, approximately 40 cm above the bench top. If a no-bounce pad is available, place it under the coil to soften the fall of the magnet/s.
3. Connect the voltage sensor to the coil.
4. What do you expect the voltage–time graph to look like when the magnet is dropped through the coil? Before continuing, sketch your prediction in the grid supplied in the Data and analysis section.
5. Hold the magnet just above the opening of the coil, start the data-collection system and then drop the magnet through the coil. Stop collecting immediately after.
6. Replace the coil with the 400-turn coil (or the coil with the next most turns) and then repeat the data collection. Do the same for each coil.
7. Display the data on one graph, annotating each to identify the number of turns. Print out and paste the graph in the Data and analysis section or sketch the graphs in the space provided.

Part B: Varying the number of magnets

Use the same set-up for this part as in Part A, but use only the 200-turn coil (or the smallest number of turns available).

1. Hold the tube containing one magnet just above the coil opening. Start the data-collection system and then drop the magnet through the coil. Stop collecting immediately after.

2 Repeat with tubes containing two, three, four and five magnets as available.

3 Display the data runs on the one graph, annotating each to identify the number of magnets. Print out and paste the graph in the Data and analysis section or sketch the graphs in the space provided.

Part C: Varying the rate of change of flux

Use the same set-up for this part as in Part A, but using only the 200-turn coil (or the smallest number of turns available) and a tube with a single magnet.

1 Roll up a piece of paper into a tube, and tape it securely. The tube should be wide enough to allow your magnet to pass through freely, but narrow enough to guide the magnet through the coil.

2 Mark four equally spaced positions on the tube and slide the tube into the coil so that the first mark is showing just above the coil opening.

3 Hold the magnet just above the tube opening. Start the data-collection system and then drop the magnet through the coil. Stop collecting immediately after.

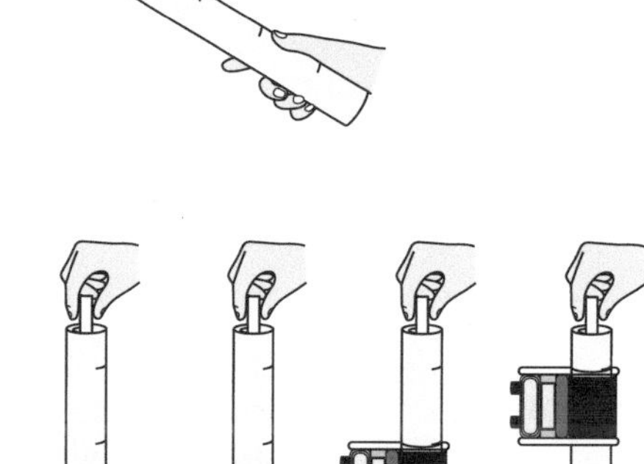

4 Slide the paper tube down into the coil until the next mark on the tube is just above the opening of the coil and repeat the data collection. Repeat the process for the remaining positions marked on the paper tube.

5 Display the data on one graph, annotating each to identify the height from which the magnets were dropped. Print out and paste the graph in the Data and analysis section or sketch the graphs in the space provided.

DATA AND ANALYSIS

Part A

1 Sketch your prediction of the voltage versus time graph.

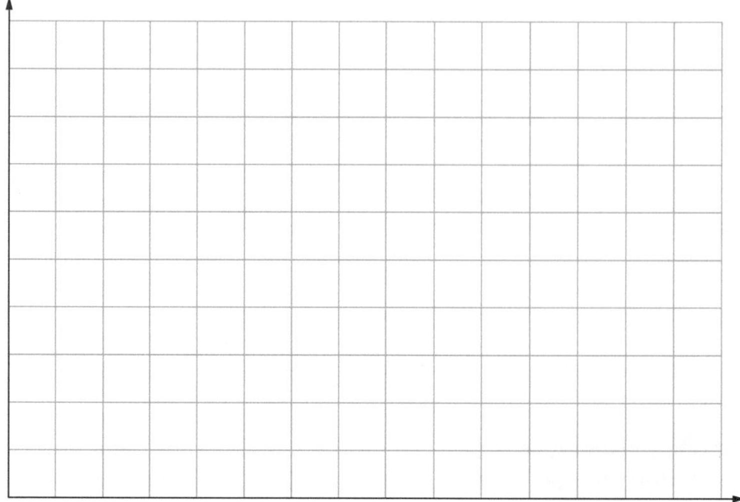

ISBN 978 1 4886 1936 6

2 If a changing magnetic field causes charges to move in a conductor, and charges moving in a conductor give rise to a magnetic field, what will be the orientation of the induced magnetic field relative to the original changing magnet field?

3 Sketch or paste your graph of the induced voltage versus time for the different number of turns in the coils.

Part B

Sketch or paste your graph of the induced voltage versus time for a different number of magnets.

Part C

Sketch or paste your graph of the induced voltage versus time for each of the different drop heights.

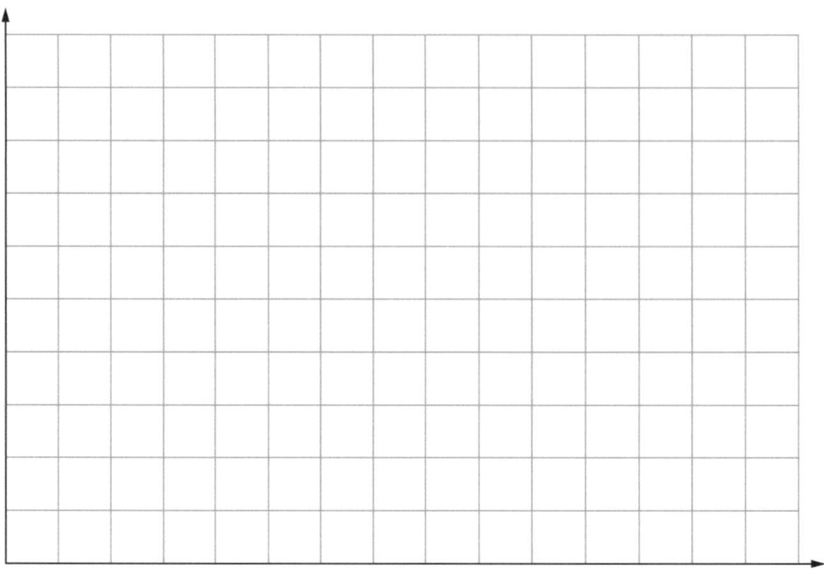

CONCLUSION

1 How did your prediction compare to the actual voltage versus time graphs?

2 Describe the relationship between the number of turns in the coils and the peak voltages observed.

3 Describe the relationship between the strength of the magnets used and the peak voltages observed. What could be measured to give a more precise relationship?

4 Describe the relationship between the height above the coil from which the magnet fell and the peak voltages observed. What could be measured to give a more precise relationship?

5 The second peak of the voltage curve is always in the opposite direction to the first peak and is also a slightly larger peak. Describe why this is the case.

RATING MY LEARNING	My understanding improved	Not confident ◄──► Very confident ○ ○ ○ ○ ○	I answered questions without help	Not confident ◄──► Very confident ○ ○ ○ ○ ○	I corrected my errors without help	Not confident ◄──► Very confident ○ ○ ○ ○ ○

 ISBN 978 1 4886 1936 6

PRACTICAL ACTIVITY 6.5

Transformer operation

Suggested duration: 45 minutes

INTRODUCTION

Transformers are an intrinsic part of any electricity supply system, whether it be solar, wind or hydro. Although a detailed analysis of transformer operation is complex, the basic idea is simple enough. Two coils are wound on one iron core so that the magnetic flux generated by one passes through the other. The coil connected to the AC supply is referred to as the primary, and the coil connected to the 'load' is the secondary. The transformer operates on the principle that, whenever a changing magnetic flux passes through a coil, there will be an induced emf. In a transformer, the changing flux originates from an alternating current in the primary coil. Because this flux also goes through the secondary coil, there will be an induced emf in both coils. In this activity the principals of operation of transformers will be investigated.

MATERIALS

- U-shaped iron cores
- pre-wound coils of different turns
- electrical leads
- low-voltage AC/DC power supply*
- two voltmeters or voltage sensors and data-collection system
- signal generator with sine and square waveform outputs

PURPOSE

To investigate the operation and the volts-to-turns ratio of a transformer.

*Ensure your DC power supply is either a true DC source such as a battery or is fully rectified and filtered. A half-wave only rectified power supply without filtering, such as standard junior laboratory supplies, will give potentially misleading results.

PROCEDURE

1 Set up the coils and core as shown in the figure below. In the diagram, the coil to the left will be referred to as the primary coil, and the one to the right will be the secondary coil. An alternating current is being applied to the primary coil and the output is being measured at the secondary.

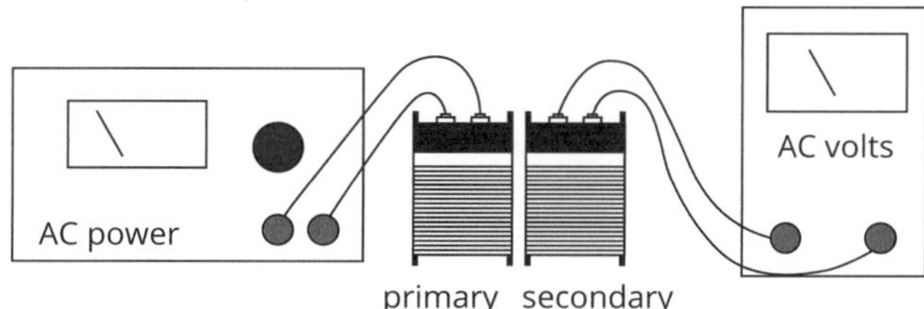

2 Set up the voltmeters or the data-collection system and connect the two voltage sensors. Set the sample rate to at least 200 Hz and display a voltage vs time graph for each sensor and a digital display of each voltage.

3 Using two coils of the same number of turns for both primary and secondary, adjust the input voltage to 6 V AC. Connect a voltmeter or voltage sensor across the primary coil and a second one across the secondary coil. Measure the voltages and record the results in Table 1 in the Data and analysis section.

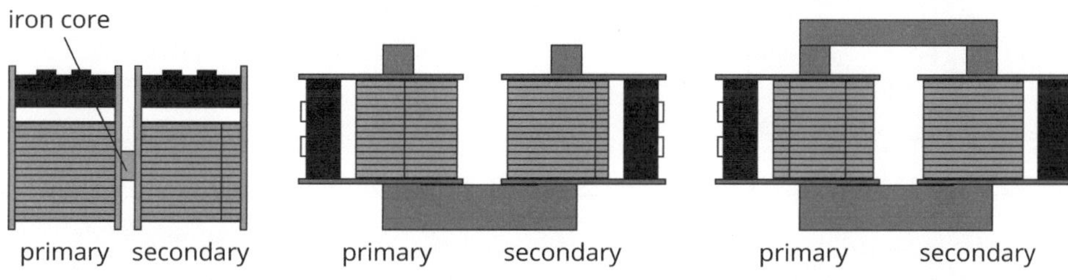

4 Insert a straight iron crosspiece between the two coils. Measure and record the input and output voltage again, keeping the input voltage from the power supply the same. Record the results in Table 1.

5 Place the coils on the sides of an open U-shaped core. Once again, measure and record the input and output voltage, keeping the input voltage from the power supply the same. Record the results in Table 1.

6 Finally, repeat the measurement after placing the cross piece over the U-shaped core. Record the results in Table 1.

Safety warning

Always use power supplies with safety cut-outs.

Do not exceed the recommended voltage levels.

Always turn off the power supply between each test.

Do not leave power running through model transformers for extended periods of time.

ISBN 978 1 4886 1936 6

7 Using the arrangement that produced the best result for the secondary output, compare each possible combination of primary and secondary coil. Record the arrangement of the coils and the respective primary and secondary outputs for each combination.

8 Using two coils with equal numbers of turns and the core configuration with the best results, replace the AC power supply with the DC supply. Adjust the output to 6 V DC and measure the input and output voltage. Record the results in Table 2 in the Data and analysis section. If using voltage sensors, draw the graph of the output voltage.

9 Replace the DC power supply with the signal generator. Set the signal generator to a sine-wave output of approximately 6 V amplitude and a frequency of 10 Hz. Record the voltage display for both the input and output voltages on the same graph of voltage versus time. Print out and paste the graph in the Data and analysis section or sketch the graphs in the space provided.

10 Finally, switch the output of the signal generator to a square wave. Keep the amplitude and the frequency the same as used with the sine wave. Record the voltage display for both the input and output voltages on the same graph of voltage versus time. Print out and paste the graph in the Data and analysis section or sketch the graphs in the space provided.

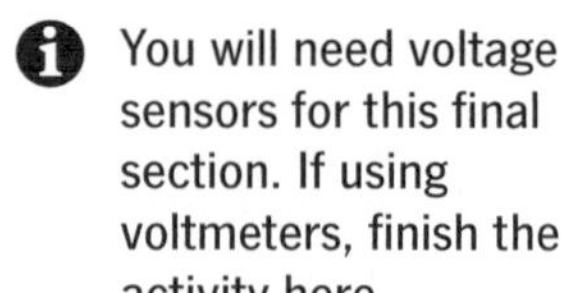
You will need voltage sensors for this final section. If using voltmeters, finish the activity here.

DATA AND ANALYSIS

TABLE 1 Transformer operation

Number of turns						
Primary (N_p)	**Secondary (N_s)**	**Input (V_p)**	**Output (V_s)**	**Ratio $\frac{N_p}{N_s}$**	**Ratio $\frac{V_p}{V_s}$**	**Core configuration**
						no core
						straight iron core
						U-shaped core
						U-shaped core with cross bar

1 Based on your results, which arrangement induces the maximum emf in the secondary coil? Why?

TABLE 2 DC input/output voltages

DC power-supply input voltage (V)	
Recorded output voltage (V)	

ISBN 978 1 4886 1936 6

PRACTICAL ACTIVITY 6.5

2 Sketch the graph of output voltage for the DC power supply.

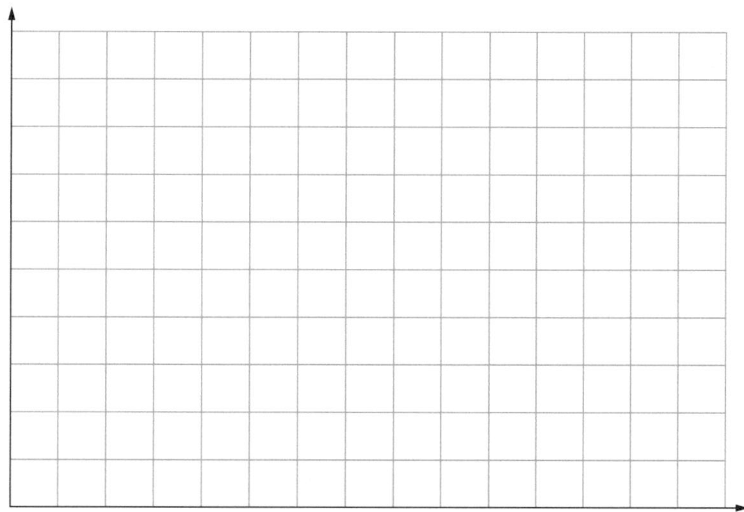

3 Sketch or paste your graph of input and output voltages for the sine-wave supply.

4 Sketch or paste your graph of the input and output voltages for the square-wave supply.

CONCLUSION

1 Compare the ratio of the number of turns $N_p : N_s$ to the ratio of the input and output voltage $V_p : V_s$. Based on your results, is the transformer an ideal transformer? Explain your conclusion.

2 Why was the voltage of the AC power supply measured at the primary coil? How does this assist the reliability of your conclusions?

3 Under what primary voltage supply conditions will a transformer work? Explain your answer with reference to the results and graphs from each input.

4 The induced voltage in the secondary coil varied depending on the configuration of the cores. Explain why this occurred.

5 A major cause of energy loss in practical transformers is eddy currents in the iron core. By what means can these be reduced?

RATING MY LEARNING	My understanding improved	Not confident ← → Very confident ○ ○ ○ ○ ○	I answered questions without help	Not confident ← → Very confident ○ ○ ○ ○ ○	I corrected my errors without help	Not confident ← → Very confident ○ ○ ○ ○ ○

 ISBN 978 1 4886 1936 6

PRACTICAL ACTIVITY 6.6

Electricity from a DC motor

Suggested duration: 20 minutes

INTRODUCTION

In this activity, the basic principle of electric power generation will be investigated. A rotating-coil generator is fundamentally just an electric motor being manually turned to produce a current.

PURPOSE

To investigate the output of a DC motor driven mechanically.

MATERIALS

- centre-zero ammeter or galvanometer (or current and voltage sensors and data-collection system)
- CRO (cathode ray oscilloscope, if not using an electronic measure)
- electric motor
- electrical leads with alligator clips or banana plugs
- gear system—use gears and a handle to make it easy to turn the shaft of the electric motor

PROCEDURE

1 Connect a small electric motor to a gear system so that the centre shaft of the motor can be manually turned. Connect the leads of the motor to a centre-zero ammeter, or a current sensor. Set up the current sensor following the manufacturer's instructions.

2 Start turning the motor (and start recording data if using an electronic measure) as slowly and as smoothly as possible. Observe the output as shown on the galvanometer or the graph of the voltage sensor. Increase the speed of rotation gradually and record your observations of the effect on the output.

3 If you are using an ammeter, stop turning the motor after a little while and replace it with a CRO. Observe the output from the motor on the oscilloscope screen. Sketch two waveforms corresponding to two different speeds of rotation on the same axes. Label the waveform corresponding to the highest speed of rotation.

DATA AND ANALYSIS

Graph of voltage versus time for two speeds of rotation

CONCLUSION

1 Is the electric motor an AC or DC generator? Explain with reference to your graph.

2 How does the speed of rotation affect the frequency and amplitude of the output voltage? Why is this the case?

RATING MY LEARNING	My understanding improved	Not confident ⟷ Very confident ○ ○ ○ ○ ○	I answered questions without help	Not confident ⟷ Very confident ○ ○ ○ ○ ○	I corrected my errors without help	Not confident ⟷ Very confident ○ ○ ○ ○ ○

DEPTH STUDY 6.1

Electric power—issues in supply and distribution

Suggested duration: 3.75–4 hours

INTRODUCTION

Electricity is an indispensable commodity in the modern world. If you don't think so, just recall the last time you had a blackout at home. The reliable and economical provision of electricity is an issue which is rarely out of the public eye and the ongoing efforts to adapt to our changing climate regularly bring new problems.

The story of the provision of electric power started with street lighting around 1880 and quickly spread to power for domestic, industrial and transport applications in the following decades. Today, the provision of electric power continues to be of major importance across the planet, with electricity being generated from a wide variety of resources and transmitted via a range of different systems.

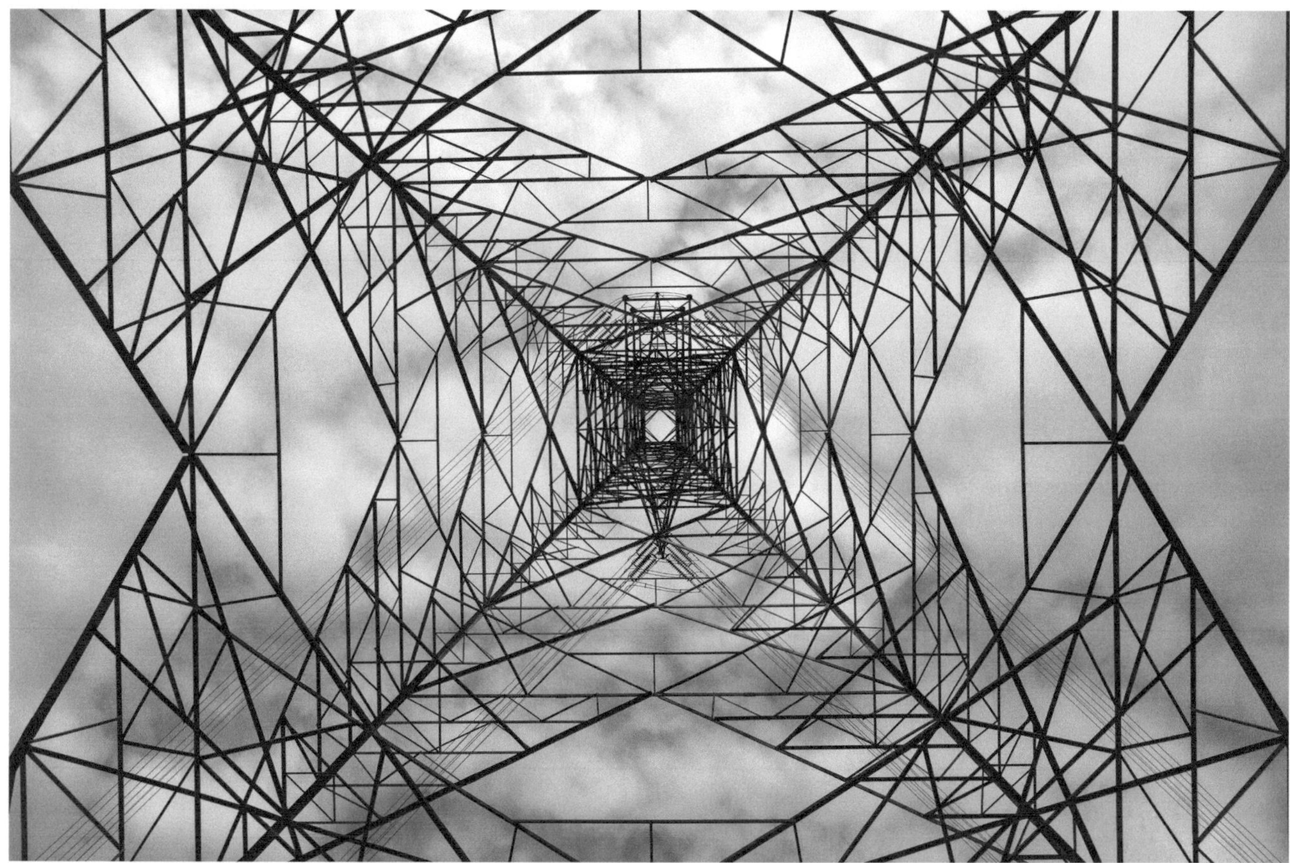

Your task is to conduct some detailed research into an aspect of the provision of electric power and its history. You will need to focus on the physics behind the issue, not just the human endeavours. A list of stimulus topics is provided here to help you get started, but don't limit yourself to these.

- The War of Currents: Edison vs Tesla (or AC vs DC), General Electric vs Westinghouse
- Generation: Spinning generators or solar? Are there other ways that electricity could be generated on a large scale?
- Transmission over long distances: AC vs DC—a decided issue?
- Transmission overground, underground, undersea, or through space?
- Matching voltages and frequencies between different sources and grids
- High-voltage switching, and three-phase power
- Sunspots and electric transmission grids

TOPIC REQUIREMENTS AND CONSTRAINTS

Your research must be conducted individually. The topic chosen must consist of an issue in the area of AC electricity generation and supply as the primary area of research. The topic must allow for the development of an answer to one clear inquiry question. A topic that has a broad range of resources available is likely to prove better than one where the resources are limited.

 ISBN 978 1 4886 1936 6

QUESTIONING AND PREDICTING

1 Carefully define your topic and define its limits. Perhaps outline what it does *not* cover.

2 Phrase your topic as a question or as a number of related questions. For example, 'What's so good about alternating current?', 'How is power shared between different Australian states?', 'How can batteries help balance the demand?' It is important to look at different sources of information before you settle on an area or question. Many electrical utility companies provide detailed information.

3 Do some initial research around answering your initial questions. Rephrase your questions to make them as clear and specific as possible. For example, 'How does DC get converted into AC?', 'Why is three-phase power so efficient in generation and transmission?', 'How is the output of a wind turbine synchronised to the grid?', 'Why is switching off 500 kV so tricky?'

4 Look back at your initial topic question. Construct a hypothesis that can be applied to answer your question. This becomes your working hypothesis and should summarise the answer to your main research question. It will most likely change after some further research.

PLANNING YOUR INVESTIGATION

5 Look back at your original question. Does it need rephrasing in light of your research? Restate your question as you will present it.

ISBN 978 1 4886 1936 6

6 Highlight the physics in your topic. Can you put it into terms your fellow students will understand?

CONDUCTING YOUR INVESTIGATION

7 Working in a small group, evaluate other students' research topics. What are the strengths of their research questions? How could you improve your own research question?

8 Carry out the remainder of your research, summarising the key points as you proceed so that your topic has a clear focus and structure.

ANALYSING DATA AND INFORMATION

Evaluate your research using the following questions.

9 Were you successful in answering your original research question? Did you need to rephrase it?

10 Do your research sources all agree? Are there any dissenting claims or explanations?

COMMUNICATING

Communicate your findings in the form of an individual documentary, media report or other short visual presentation.

11 Did your presentation generate questions from your audience?

 ISBN 978 1 4886 1936 6

Multiple choice

1 Which two of the following statements explain why high voltages should be used in electricity transmission?

A High transmission voltages require high current.

B For a given power, the larger the transmission voltage the smaller the current.

C High transmission voltage causes less radiation of electric energy.

D To minimise power loss in the wires, the current must be small.

2 In a particle accelerator, electrons are travelling along the path indicated. They are made to change direction by a magnetic field.

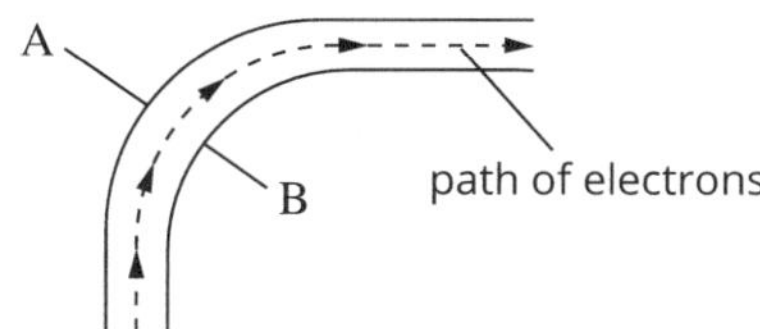

Which one of the following best gives the direction that the magnetic field must have to keep the electrons in this curved path?

A direction A on the diagram

B direction B on the diagram

C out of the page

D into the page

3 In a DC electric motor, the magnetic field is created by a pair of permanent magnets and an armature coil mounted in the field so that it can rotate freely. The current to the coil is supplied via a commutator. The function of the commutator is to:

A keep the direction of the current in the coil the same at all times.

B reverse the direction of the coil current once in every rotation.

C reverse the direction of the coil current twice in every rotation.

D reverse the direction of the coil current four times in every rotation.

4 A pair of straight parallel wires carrying currents of I and $2I$ is set up a distance d apart. They experience a force of F between them. What force will act if both currents are doubled *and* the distance d is halved?

A F

B $2F$

C $4F$

D $8F$

E $16F$

5 An AC generator produces an alternating signal when rotating at a certain frequency. If rotated twice as fast, its:

A period and amplitude will both double.

B amplitude will double but its period will be halved.

C period and amplitude will both be halved.

D period will double but its amplitude will be halved.

6 Which of the following changes will *not* improve the performance of a transformer?

A increasing the number of turns on both primary and secondary sides

B introducing a laminated iron core in place of a solid core

C installing cooling fans around the body of the transformer

D increasing the flux linkage between the primary and secondary sides

Short answer

7 A 20 m horizontal wire carries a DC current of 60 A from east to west across a street. The Earth's magnetic field (5.0×10^{-5} T) points north but is inclined upwards at 25° to the horizontal.

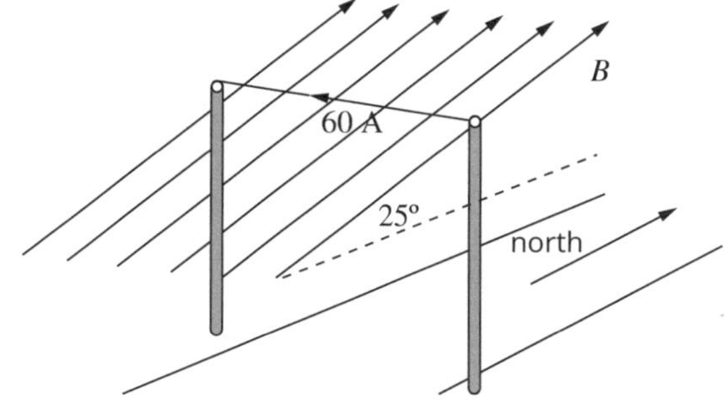

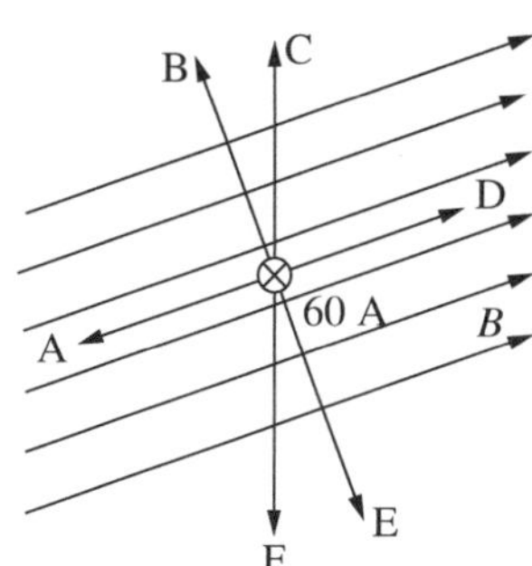

a Calculate the size of the magnetic force acting on the wire.

b The lower diagram shows the view of the wire and field looking from the east towards the west. Which one of the arrows best indicates the direction of the force on the wire?

c What would happen to the wire if instead of DC the current was 60 A AC?

8 A square coil of 50 turns whose side is 3.0 cm long is moving at $12\,\text{cm}\,\text{s}^{-1}$ into a uniform magnetic field of 4.0×10^{-4} T. The coil has a resistance of 6.0 Ω (represented by a single resistor symbol). At the instant shown it is halfway into the field.

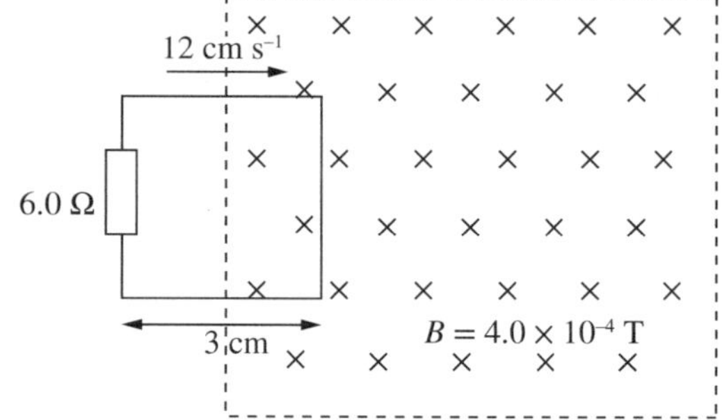

a How much flux is threading the coil at the instant shown?

b What is the size of the emf generated in the coil?

c Describe and calculate the forces, if any, acting on the coil.

9 In the situation depicted below, some students connect a 6.0 V battery to the primary winding of a transformer. The secondary winding is attached to a 400 Ω resistor and an AC ammeter.

iron core
I_p I_s
6.0 V V_p V_s 400 Ω A
200 turns 1000 turns

a In a practical transformer, the core is laminated. Why?

b After a few minutes, the students notice that the transformer is getting very hot. They were also expecting to find a secondary current of 75 mA but do not see any current reading at all. Explain their predictions and why they were wrong.

10 a What is the electromagnetic definition of the ampere?

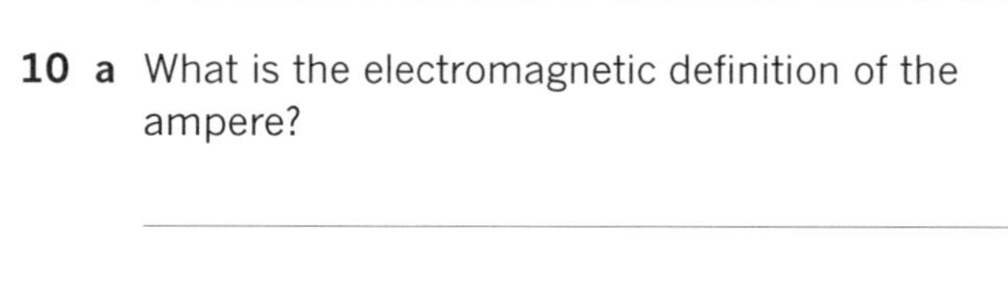

b A straight vertical wire carrying a current of 7.00 A in an upwards direction is placed in an area where the Earth's field has a magnitude of 5.40×10^{-5} T and points north with an angle of 35° below the horizontal. Determine the size and direction of the magnetic force on 2.00 m of the wire.

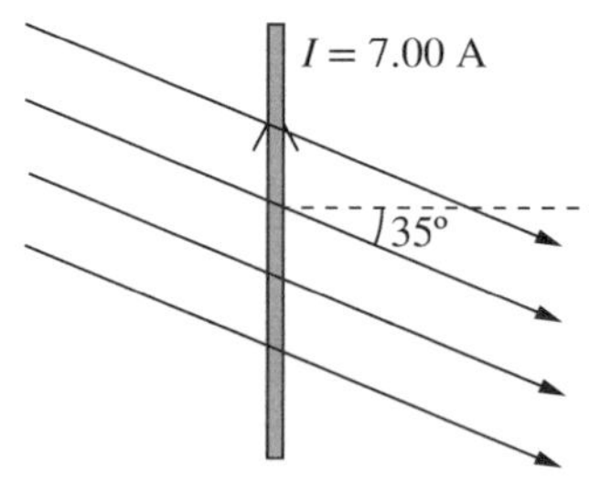

c A second vertical wire carrying a current of 2.60 A in the opposite direction is placed 1.50 cm from the first wire. Ignoring the effect of the Earth's field, determine the magnitude and direction of the force acting per metre between the wires. ($\mu_0 = 4\pi \times 10^{-7}\,\text{T}\,\text{m}\,\text{A}^{-1}$).

ISBN 978 1 4886 1936 6

11 A tiny charged plastic sphere of mass 4.50 mg is held stationary halfway between two horizontal, charged parallel plates. The plates, inside a vacuum chamber, are 8.00 mm apart and the potential difference between them is 400 V.

a What is the electric field strength between the plates?

b What is the size of the charge on the sphere?

c The potential difference between the plates is now doubled.

i Will the sphere move?

ii If so, with what acceleration?

d The plates are now quickly rotated through 90°.

i Will the sphere move?

ii If so, with what acceleration?

12 A stream of H^+ ions are being injected at a speed of $8.0 \times 10^4 \, m\,s^{-1}$ into the space between two *uncharged* parallel plates, P and Q, 2.0 cm apart. In addition, there is a uniform magnetic field in the space which is directing the ions along path A.

P
A
H^+ gun
C
B
Q

a Inside the H^+ gun, the hydrogen ions are being accelerated by a potential difference. If each hydrogen ion has a mass of 1.67×10^{-27} kg and a charge of 1.60×10^{-19} C, what is the size of the accelerating voltage?

b What is the direction of the magnetic field?

c A voltage of 5.0 kV is now applied to the plates such that the ions travel undeviated along path C. Which plate is positive?

d Calculate the strength of the magnetic field.

e Describe what would happen if:

i the plates were moved closer together

ii He^{2+} ions were substituted for the hydrogen ions in the gun with the same accelerating voltage. (Each He^{2+} ion has twice the charge and four times the mass of an H^+ ion.)

MODULE

7 The nature of light

Outcomes

By the end of this module you will be able to:

- develop and evaluate questions and hypotheses for scientific investigation PH12-1
- design and evaluate investigations in order to obtain primary and secondary data and information PH12-2
- conduct investigations to collect valid and reliable primary and secondary data and information PH12-3
- select and process appropriate qualitative and quantitative data and information using a range of appropriate media PH12-4
- communicate scientific understanding using suitable language and terminology for a specific audience or purpose PH12-7
- describe and analyse evidence for the properties of light and evaluate the implications of this evidence for modern theories of physics in the contemporary world PH12-14.

Content

ELECTROMAGNETIC SPECTRUM

INQUIRY QUESTION What is light?

By the end of this module you will be able to:

- investigate Maxwell's contribution to the classical theory of electromagnetism, including:
  - unification of electricity and magnetism
  - prediction of electromagnetic waves
  - prediction of velocity (ACSPH113) CCT ICT
- describe the production and propagation of electromagnetic waves and relate these processes qualitatively to the predictions made by Maxwell's electromagnetic theory (ACSPH112, ACSPH113)
- conduct investigations of historical and contemporary methods used to determine the speed of light and its current relationship to the measurement of time and distance (ACSPH082) CCT ICT
- conduct an investigation to examine a variety of spectra produced by discharge tubes, reflected sunlight or incandescent filaments
- investigate how spectroscopy can be used to provide information about: ICT
  - the identification of elements
- investigate how the spectra of stars can provide information on: CCT ICT
  - surface temperature
  - rotational and translational velocity
  - density
  - chemical composition.

LIGHT: WAVE MODEL

INQUIRY QUESTION **What evidence supports the classical wave model of light and what predictions can be made using this model?**

By the end of this module you will be able to:

- conduct investigations to analyse qualitatively the diffraction of light (ACSPH048, ACSPH076) ICT
- conduct investigations to analyse quantitatively the interference of light using double-slit apparatus and diffraction gratings $d \sin\theta = m\lambda$ (ACSPH116, ACSPH117, ACSPH140) ICT N
- analyse the experimental evidence that supported the models of light that were proposed by Newton and Huygens (ACSPH050, ACSPH118, ACSPH123) CCT
- conduct investigations quantitatively using the relationship of Malus' law $I = I_{max}\cos^2\theta$ for plane polarisation of light, to evaluate the significance of polarisation in developing a model for light (ACSPH050, ACSPH076, ACSPH120). ICT N

LIGHT: QUANTUM MODEL

INQUIRY QUESTION **What evidence supports the particle model of light and what are the implications of this evidence for the development of the quantum model of light?**

By the end of this module you will be able to:

- analyse the experimental evidence gathered about black-body radiation, including Wien's law, related to Planck's contribution to a changed model of light (ACSPH137) CCT ICT N
  - $\lambda_{max} = \frac{b}{T}$
- investigate the evidence from photoelectric effect investigations that demonstrated inconsistency with the wave model for light (ACSPH087, ACSPH123, ACSPH137) CCT ICT
- analyse the photoelectric effect $K_{max} = hf - \phi$ as it occurs in metallic elements by applying the law of conservation of energy and the photon model of light (ACSPH119). ICT N

LIGHT AND SPECIAL RELATIVITY

INQUIRY QUESTION **How does the behaviour of light affect concepts of time, space and matter?**

By the end of this module you will be able to:

- analyse and evaluate the evidence confirming or denying Einstein's two postulates:
  - the speed of light in a vacuum is an absolute constant
  - all inertial frames of reference are equivalent (ACSPH131)

Module 7 • The nature of light

- investigate the evidence, from Einstein's thought experiments and subsequent experimental validation, for time dilation $t = \frac{t_0}{\sqrt{\left(1 - \frac{v^2}{c^2}\right)}}$ and length contraction $l = l_0\sqrt{1-\frac{v^2}{c^2}}$, and analyse quantitatively situations in which these are observed, for example:
  - observations of cosmic-origin muons at the Earth's surface ICT N
  - atomic clocks (Hafele–Keating experiment) CCT ICT N
  - evidence from particle accelerators CCT ICT N
  - evidence from cosmological studies ICT
- describe the consequences and applications of relativistic momentum with reference to:
  - $p_v = \frac{m_0 v}{\sqrt{\left(1-\frac{v^2}{c^2}\right)}}$ ICT N
  - the limitation on the maximum velocity of a particle imposed by special relativity (ACSPH133) CCT
- use Einstein's mass–energy equivalence relationship $E = mc^2$ to calculate the energy released by processes in which mass is converted to energy, for example: (ACSPH134) ICT N
  - production of energy by the Sun
  - particle–antiparticle interactions, e.g. positron–electron annihilation
  - combustion of conventional fuel.

Key knowledge

Electromagnetic spectrum

ELECTROMAGNETISM

Light exhibits behaviours that are characteristic of both waves and particles. The wave characteristics of light cannot solely be modelled as a mechanical wave because light can travel through a vacuum. The physicist James Maxwell unified the theories of electricity and magnetism to find that light is a form of **electromagnetic radiation** (EMR). Recall that a point charge generates an electric field and moving point charges—a current—generate a magnetic field. Maxwell put these two ideas together, and stated that this would produce two mutually propagating fields, as shown in Figure 7.1, known as EMR or electromagnetic (EM) waves.

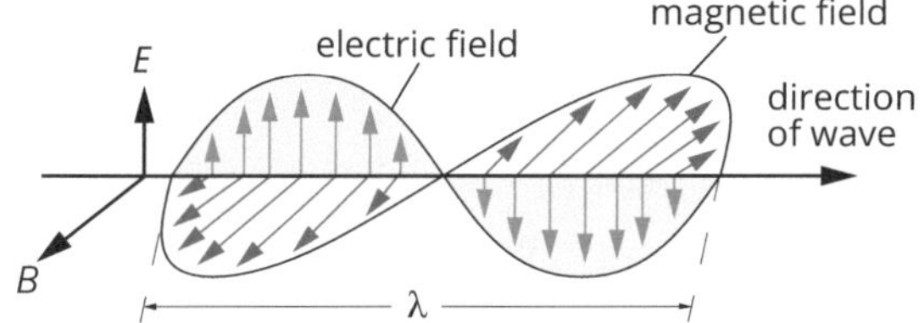

FIGURE 7.1 Electromagnetic radiation

EM waves are transverse waves made up of mutually perpendicular, oscillating electric and magnetic fields. These fields are perpendicular to the direction of propagation of the radiation, and the frequency, and therefore the wavelength λ, of both fields is the same, as can be seen in Figure 7.1. EM waves cause charges to oscillate at the frequency of the wave. This also works in reverse. Oscillating charges produce EM waves of the same frequency as the oscillation.

The wave equation ($v = f\lambda$) can be used to calculate the frequency and wavelength of EM waves. The speed they travel is the speed of light.

GO TO ➤ Year 11 Module 3

Maxwell's theoretical calculations for the speed of EMR closely matched the experimentally determined values for the speed of light—so closely that it provided a clue that light must be a form of EMR. (Light travels through a vacuum at approximately $c = 3.0 \times 10^8\,\text{m s}^{-1}$, which is so reliable it is a fixed SI unit.) The wave equation can therefore be changed to:

$c = f\lambda$

where:

c is the speed of light (m s^{-1})

f is the frequency of the wave (s^{-1} or Hz)

λ is the wavelength of the wave (m).

EMR can be used for a variety of purposes depending on the properties of the waves, which are determined by their frequency. The **EM spectrum** is the range of all types of EMR, and is divided into 'bands' according to how the types of EMR are used (Figure 7.2). The shorter the wavelength of the EM wave, the greater its penetrating power. The most penetrating waves are gamma rays. On the other hand, long wavelength waves such as AM radio waves have such low penetrating power that they cannot even escape Earth's atmosphere.

SPECTROSCOPY

Spectroscopy is the branch of science which investigates the **spectra** produced when matter interacts with or emits EMR. A **spectroscope** is the device used to determine the spectra and works by combining a diffraction grating with a viewing telescope or digital converter.

A diffraction grating is used to separate the different colours of light. It acts as a super prism.

When a graduated scale is added to the spectroscope to make a **spectrometer**, the specific wavelengths in the spectrum can be determined. The spectroscope determines the chemical makeup of a visible source of light because each element has its own distinctive **line spectrum** or **emission spectrum**.

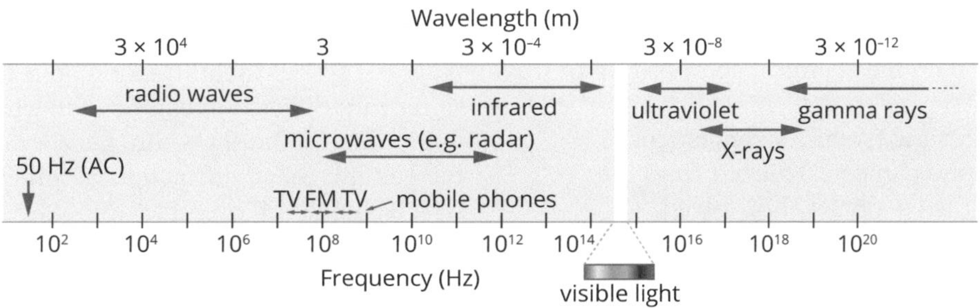

FIGURE 7.2 Electromagnetic spectrum

An emission spectrum (Figure 7.3b) shows only certain colours or specific wavelengths of light. When elements are heated to high temperatures or have an electrical current passed through them, they produce light. Atoms within the material absorb energy and become 'excited', and this makes the atom unstable. Eventually the atom will return to the 'unexcited' or **ground state**, and when this happens, the energy that had been absorbed is released. The wavelength of that energy (now known as a photon, which will be discussed later in this module) will depend on the amount of energy released.

In 1819, German physicist Joseph von Fraunhofer reported many dark lines appearing in the spectrum of sunlight. Gustav Kirchhoff and Robert Bunsen, 50 years later, deduced that the dark lines were due to these colours (wavelengths or frequencies) being absorbed by gases as light made its way through the outer atmosphere of the Sun. This is called an absorption spectrum and is shown in Figure 7.3a.

By comparing the emission spectrum to the absorption spectrum viewed using a spectroscope, we can learn about the chemical makeup of an object. A line emission spectrum is produced by energised atoms, while an absorption spectrum is created when white light passes through a cold gas. The absorption and emission spectra for hydrogen are shown in Figure 7.3.

FIGURE 7.3 (a) Absorption spectrum of hydrogen, and (b) emission spectrum of hydrogen

The emission and absorption spectra of hydrogen were of interest to scientists as it had been recognised that lines in the absorption spectrum of hydrogen matched lines in the solar spectrum. This absorption spectrum allowed astronomers to determine that the Sun is largely composed of hydrogen and smaller amounts of other elements. Interpreting the visible spectrum of stars further revealed the stars' temperature, composition, age and rotation.

In spectroscopic analysis, the wavelength of absorption (or emission) lines and their intensity provides information on the chemical composition of the material. The frequency of energy emitted or absorbed by an atom represents the specific arrangement of energies permitted and can be calculated from the difference between the energy levels involved:

$$\text{i.e. } E_2 - E_1 = hf = \frac{hc}{\lambda}$$

where:

h is Planck's constant (6.262×10^{-34} J s)

f is the frequency of the light (Hz)

c is the speed of light ($3.0 \times 10^8 \text{ m s}^{-1}$)

λ is the wavelength of the light (m).

Planck's constant is also sometimes given, using electron-volts, as $h = 4.136 \times 10^{-15}$ eV s.

Light: Wave model

DIFFRACTION AND INTERFERENCE

As mentioned above, light exhibits behaviours of both a wave and a particle. This section will focus on the characteristics of light that make it act like a wave. A wave model explains a wide range of light-related phenomena, including diffraction and interference.

The theoretical basis for wave propagation in two dimensions was first explained by Christiaan Huygens. Huygens' principle states that each point on a wavefront can be considered as a source of secondary wavelets (i.e. small waves). These wavelets combine to produce a new plane wavefront, as shown in Figure 7.4. Circular waves are propagated in a similar way.

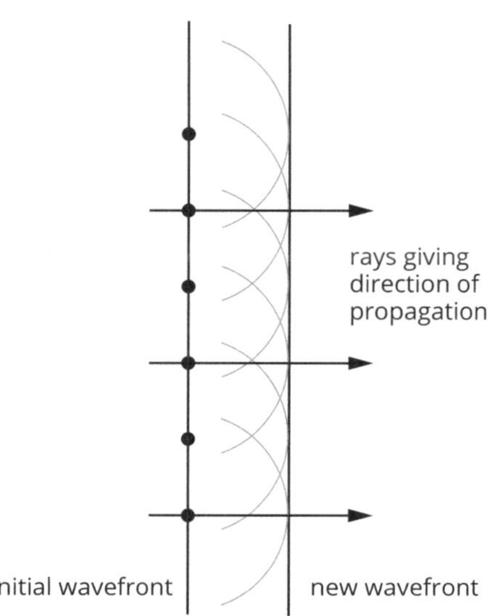

FIGURE 7.4 Huygens' principle—plane wavefront

Diffraction is the concept that is explained using Huygens's principle, and is defined as the bending of a wave around the edges of an opening or an obstacle. When a plane (straight) wave passes through a narrow opening (a slit) or meets a sharp object, it experiences diffraction. Significant diffraction occurs when the wavelength of the wave is similar to, or larger than, the size of the diffracting object. In Figure 7.5a, significant diffraction can be seen when the wavelength approximates the slit width, i.e. $\lambda \approx w$. In Figure 7.5b, as the gap increases, diffraction becomes less obvious, since $\lambda \ll w$, but is still present.

ISBN 978 1 4886 1936 6

(a)

(b)

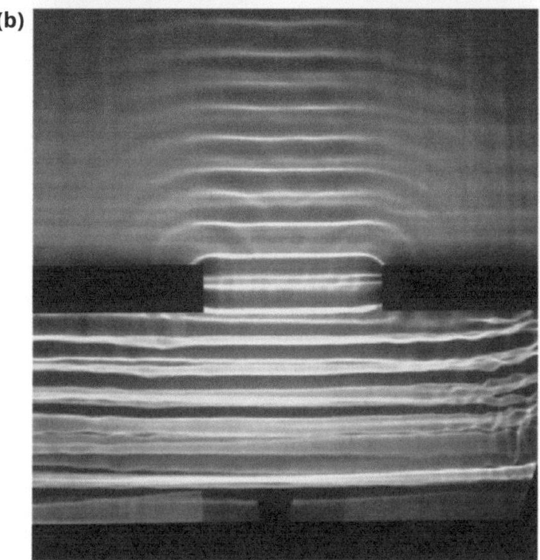

FIGURE 7.5 The diffraction of water waves in a ripple tank

In 1803, Young performed an experiment in which he shone monochromatic light on a screen containing two very tiny slits. **Monochromatic** light is light of a single colour, which means it has a single wavelength and frequency; i.e. it is coherent.

According to the particle theory, light should have passed directly through the slits to produce two bright lines or bands on the screen (Figure 7.6a). Instead, Young observed a series of bright and dark bands or 'fringes' (Figure 7.6b), which you would expect from diffracted waves. Young's double-slit interference experiment provided evidence to support the wave model of light.

Light waves have very small wavelengths (typically 500 nm, although of course it changes with colour) and so do not diffract noticeably in everyday life. Usually, diffraction occurs with artificially constructed materials like CDs or commercially produced diffraction gratings. A **diffraction grating** is a piece of material that contains a large number of very closely spaced parallel gaps or slits. As a result of this, the light waves that emerge from the slits will interact. In some places the interactions will be constructive, and in other places the interactions will be destructive. **Constructive interference** occurs when the waves are of equal frequency and in phase so that the amplitudes of the waves add together. In comparison, **destructive interference** occurs when the waves are of equal frequency but are not in phase, producing an amplitude equal to the subtraction of the waves. When these light waves are made to shine on a screen, the areas of constructive interference will appear as bright bands (known as antinodal lines) and areas of destructive interference will appear as dark bands (known as nodal lines). The pattern of dark and light bands that is seen when light passes through a single small gap is called a **diffraction pattern**. When white light, which contains a number of different colours, shines through a diffraction grating, each colour is diffracted by a different amount and forms its own set of coloured fringes.

Consider a double-slit experiment with coherent light shining through two slits, S_1 and S_2, as shown in Figure 7.7. At a particular point on the screen, P, each wave train will have travelled a different distance, i.e. S_1P and S_2P. The **path difference** (pd) is the difference in the distance travelled by each wave train from a pair of slits to the same point on the screen.

In Young's experiment, it is possible to predict where the bright fringes will occur by considering the geometry of the situation. Consider two slits, S_1 and S_2, that are separated by a distance, d. If a line is drawn at right angles halfway between the two slits directly across to the screen, then any point, P, on the screen can be identified by its angle, θ, from this line (shown in Figure 7.7). Data from Young's experiment can be analysed using the formula:

path difference, pd $= d \sin\theta = m\lambda$

where:

d is the slit separation (m)

θ is the position of the point on the screen (as an angle from the perpendicular bisector between the slits)

m is any whole number, i.e. 0, 1, 2, 3...

λ is the wavelength of the light waves (m).

(a)

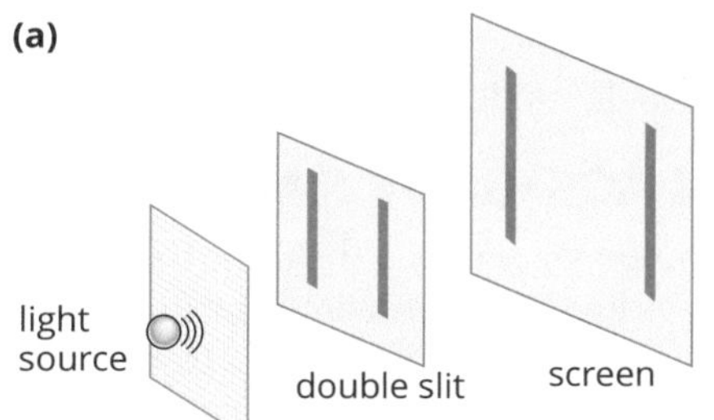

(b)

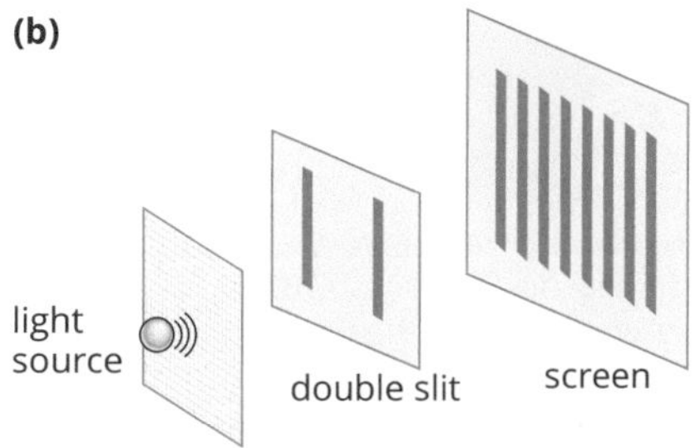

FIGURE 7.6 Young's double-slit experiment. (a) According to the particle theory of light, two bright lines should be seen on the screen. However, a series of bright and dark lines were seen (b), which supported a wave model of light.

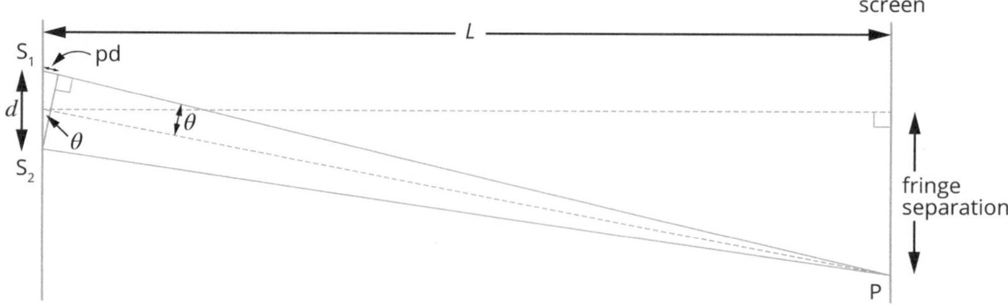

FIGURE 7.7 The geometry of two-point source interference

Constructive interference of coherent waves occurs when the path difference is equal to:

pd = $m\lambda$, where $m = 0, 1, 2, 3\ldots$

Destructive interference of coherent waves occurs when the path difference equals an odd number of half wavelengths:

$\text{pd} = \left(m - \frac{1}{2}\right)\lambda$ where $m = 1, 2, 3\ldots$

POLARISATION

Polarisation occurs when a transverse wave is only allowed to vibrate in one direction. Light produced by sources such as a light globe or the Sun is unpolarised, which means that it can be thought of as a collection of waves, each with a different plane of polarisation. Certain materials can act as polarising filters for light. These only transmit the waves or components of waves that are polarised in a particular direction; the remainder are absorbed (Figure 7.8).

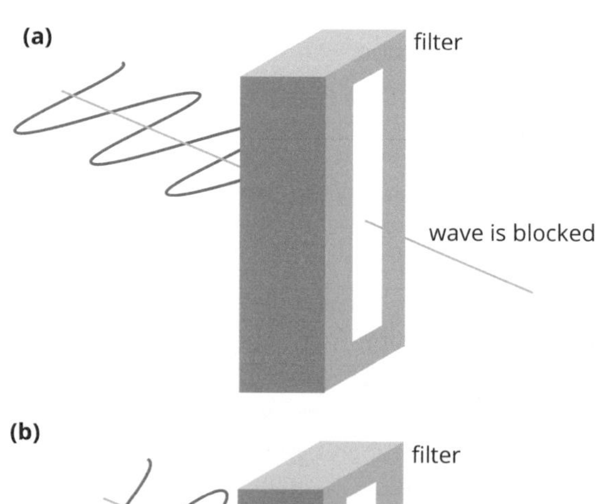

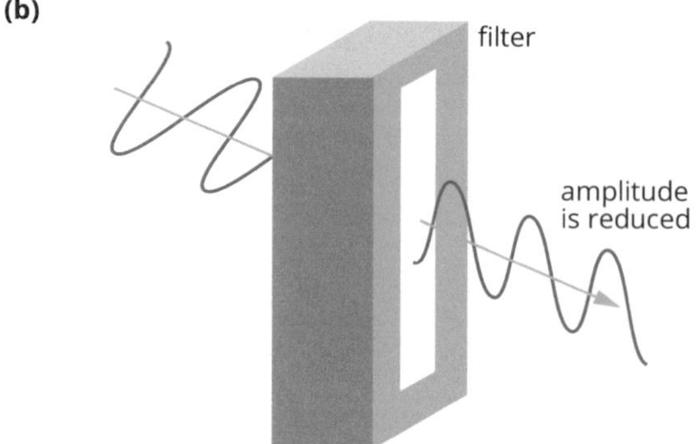

FIGURE 7.8 (a) A horizontally polarised wave cannot pass through a vertically oriented polarising filter. (b) A diagonally polarised wave has its horizontal component blocked by the vertically orientated polarising filter. A vertically polarised wave of reduced amplitude passes through it.

The formula to calculate the intensity of light after it has passed through a polarisation filter is known as Malus' law:

$I = I_{max} \cos^2 \theta$

where:

I is the intensity of light passing through the filter (cd)

I_{max} is the intensity of light entering the filter (cd)

θ is the angle between the direction of polarisation of the light entering the filter, θ_0, and the axis of polarisation of the filter, θ_1.

These values are described in Figure 7.9.

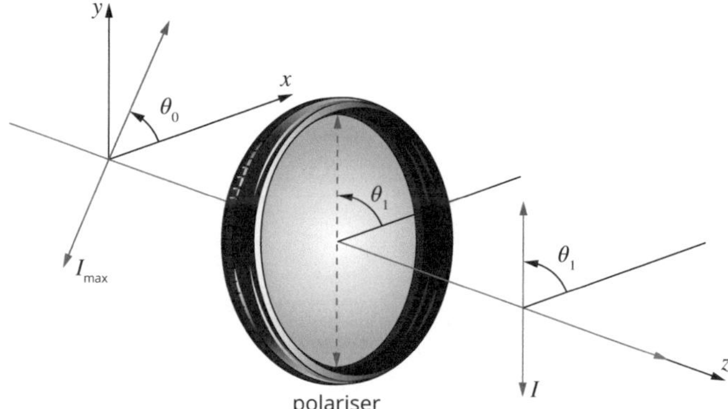

FIGURE 7.9 In Malus's law, θ is the angle between the direction of polarisation of the light entering the filter, θ_0, and the axis of polarisation of the filter, θ_1, i.e. $\theta = \theta_1 - \theta_0$.

Intensity is a measure of the energy emitted over an area, and so it can also be given in units of W m^{-2}. In Year 11 it was shown that the intensity of light propagating from a source will decrease by an inverse square relationship, so that $I \propto \frac{1}{r^2}$.

GO TO ➤ Year 11 Module 3

Light: Quantum model

So far you have explored the behaviour of light when it is acting like a wave. This section will outline the key experiments that led to the development of a 'new' particle model of light, known as the quantum model.

BLACK-BODY RADIATION

You may recall that the term EMR includes such things as visible light, ultraviolet light and infrared radiation. All forms of EMR are essentially the same, differing only in their frequency and, therefore, their wavelength. EMR is emitted by all objects whose temperature is above absolute zero (0 K or −273°C). The higher the temperature of an object, the higher the frequency and the shorter the wavelength of the emitted radiation. The peak wavelength, at which an object will emit the maximum intensity of radiation, is dependent on the object's surface temperature and is given by Wien's law. Wien's law states that:

$$\lambda_{max}T = 2.898 \times 10^{-3}\,\text{mK}$$

where:

λ_{max} is the peak wavelength of the emitted radiation (m)

T is the surface temperature of the object (K).

Wien's work on the wavelength of the radiation emitted by a hot, dense object was based initially on a theoretical object called a **black body**. A black body is a good absorber of EMR, and all light originates from the body itself (to reach thermal equilibrium)—none is reflected. An example of a black body is a tripod gauze (as it's being heated by Bunsen burner it turns red, orange and even white). Wien's law makes it possible to determine the approximate temperature of stars, assuming that they emit radiation similar to that emitted by a black body.

In 1900, the German physicist Max Planck was studying the spectrum for light emitted by hot objects. Planck and other scientists had discovered that certain features of this spectrum could not be explained using a wave model for light. Planck proposed that energy released by a black body was emitted by atoms and that these atoms could only vibrate at certain frequencies (it was later found that it wasn't atoms vibrating, it was the electrons—this will be discussed later). He had to assume that the energy released by the atoms was given off in small energy packets called **quanta**. The energy of a quantum is proportional to its frequency:

$$E = hf = \frac{hc}{\lambda}$$

where:

h is Planck's constant (6.262×10^{-34} J s)

f is the frequency of the light (Hz)

c is the speed of light ($3.0 \times 10^{8}\,\text{m s}^{-1}$)

λ is the wavelength of the light (m).

The electron-volt is an alternative (non-SI) unit of energy where 1 eV = 1.602×10^{-19} J. Planck's constant is sometimes given in eV as 4.136×10^{-15} eV s.

THE PHOTOELECTRIC EFFECT

At the start of the 20th century, another phenomenon that could not be explained using the wave model for light was being observed. Scientists noticed that when some types of EMR are incident on a piece of metal (i.e. when the light strikes the metal), the metal becomes positively charged. This positive charge is due to electrons being ejected from the surface of the metal. The electrons became known as **photoelectrons** because they were released due to incident light or other forms of EMR. The phenomenon is known as the **photoelectric effect**.

The experiment used to investigate the photoelectric effect involves illuminating a metal plate (the cathode) mounted inside a tube that has an oppositely charged end to the plate (the anode). The resulting electric field helps the photoelectrons cross the gap to the anode (positive potential). A galvanometer is used to measure the current. This is shown in Figure 7.10.

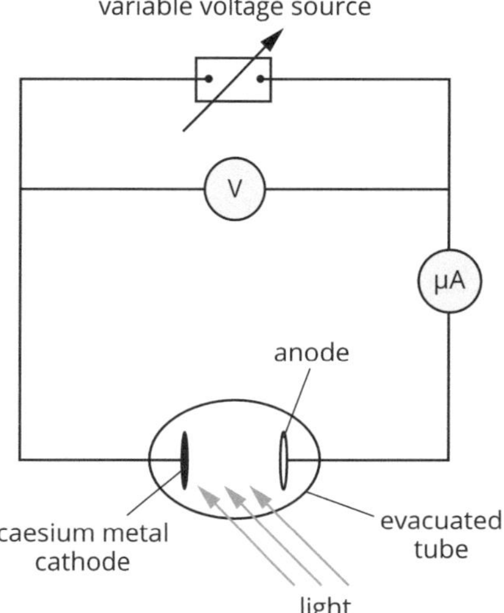

FIGURE 7.10 Circuit diagram of an experimental investigation of the photoelectric effect

German physicist Philipp Lenard discovered that, for a particular cathode metal, there is a certain frequency of light below which no photoelectrons are observed. This is called the **threshold frequency**, f_0. The photoelectric effect will only emit photoelectrons from a clean metal surface if the frequency of the incident light is greater than a threshold frequency, f_0.

If $f < f_0$, no electrons are released.

If $f > f_0$, the rate of electron release (the photocurrent) is proportional to the intensity of the light and occurs without any time delay.

Figure 7.11 is a graph of photocurrent (I) plotted as a function of the voltage (V) applied between the cathode and the anode for different light intensities. For brighter light ($I_2 > I_1$) of the same frequency ($f_1 = f_2$), there is a higher photocurrent, but the same stopping voltage, V_0. The **stopping voltage** is a voltage, V_0, at which no photoelectrons reach the collector. The stopping voltage is negative and so photoelectrons are attracted back towards the illuminated cathode and repelled by the collector electrode (anode), and the photocurrent is reduced. For a particular frequency of light incident on a particular metal, this stopping voltage is a constant. If two different frequencies of light are compared, the higher frequency light has a higher stopping voltage. This graph also shows another important property—the forward voltage does not alter the rate of electron release (i.e. the photocurrent).

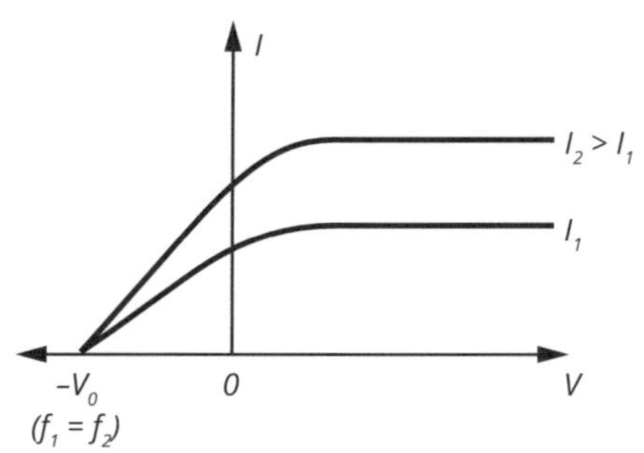

FIGURE 7.11 Photocurrent (I) plotted as a function of the voltage (V)

There are a few energy transformations involved when the light source illuminates the metal plate.

1. The energy of the incoming photon ($E = hf$). This is the total energy available in the collision.
2. The energy required to overcome the attractive forces acting on the electron in the 'sea'. This is called the **work function**, ϕ.
3. The maximum kinetic energy of electrons after they have been ejected from the surface of the metal.

By experiment, the maximum kinetic energy for the electrons, K_{max} (i.e. that of the fastest electron), can be found by using a reverse voltage, the stopping voltage, V_0. The formula is:

$$K_{max} = q_e V_0$$

where q_e is the charge on an electron (-1.602×10^{-19} C).

For a given metal, a certain amount of energy is needed to eject the electron. This is called the work function. The work function, ϕ, for the metal is given by $\phi = hf_0$, and is different for each metal. If the frequency of the incident light is greater than the threshold frequency, then a photoelectron will be ejected with some kinetic energy up to a maximum value.

The maximum kinetic energy of the photoelectrons emitted from a metal is the energy of the photons minus the work function, ϕ, of the metal: $K_{max} = hf - \phi$.

A graph of K_{max} versus frequency will have a gradient equal to Planck's constant, h, and a y-intercept equal to the work function, ϕ. Figure 7.12 shows that magnesium has a high threshold frequency and is in the ultraviolet region. The threshold frequency for potassium is in the visible region. The work function for both metals can be taken from the y-intercepts.

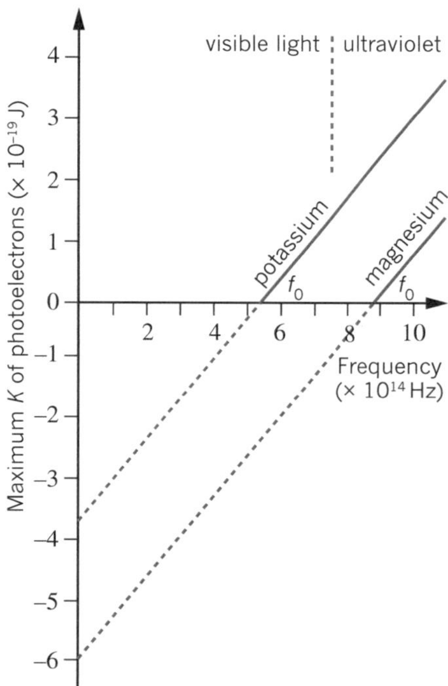

FIGURE 7.12 Kinetic energy of emitted photoelectrons versus frequency of incident light for potassium and magnesium

A number of phenomena related to the behaviour of light, such as the photoelectric effect, can only be explained using the concept of photons or light quanta. The wave approach to light could not explain various features of the photoelectric effect: the existence of a threshold frequency, the absence of a time delay when using very weak light sources, and increased intensity of light resulting in a greater rate of electron release rather than increased electron energy. According to the wave model, the frequency of light should be irrelevant to whether or not photoelectrons are ejected. Since a wave is a form of continuous energy transfer, it would be expected that energy from the wave would build up in the metal over time.

Einstein used Planck's concept of a photon to explain the photoelectric effect, stating that each electron release was due to an interaction with only one photon. The photon model of light explained the existence of a threshold frequency for each metal, the absence of a time delay for the photocurrent even for weak light sources and why brighter light resulted in a higher photocurrent.

Light exhibits wave properties in some situations and particle properties in other situations. The concept of **wave–particle duality** is used to describe the dual nature of light.

Light and special relativity

The theory of relativity had its beginning in the peculiar behaviour of light waves. This section will discuss how light is very important in special relativity.

EINSTEIN'S POSTULATES

Galileo and Newton developed theories of motion. These theories allowed the relative motion of low-speed objects to be modelled mathematically. Galileo's principle of relativity states that you cannot tell if you are moving or not, if you are travelling at constant velocity, without looking outside of your own **frame of reference**.

A frame of reference describes where an observation is being made from.

Based on the work of Galileo, Isaac Newton developed mathematical models to describe motion, and using these equations the velocity of objects can be calculated relative to any frame of reference if the velocity of the frame of reference is known. Newton had referred to these as **inertial frames of reference**, as the law of inertia applied within them.

Einstein decided that Galileo's principle of relativity was so elegant it simply had to be true, and he was also convinced that Maxwell's electromagnetic equations, and their predictions, were sound. Most physicists believed that the constant speed of light predicted by Maxwell's equations referred to the speed of light relative to a **medium** (a substance it travelled through).

ISBN 978 1 4886 1936 6

However, as light can travel through a vacuum, this medium was no ordinary material and physicists gave it the name **aether**. This was a real problem for Einstein because if the speed of light was fixed in the aether, it must depend on the velocity of an inertial frame in the aether which would be in direct conflict with the principle of Galilean relativity, which Einstein was reluctant to abandon.

Though Einstein accepted both Galileo's and Maxwell's theories despite the apparent contradiction, this still left the question: How could two observers travelling at different speeds see the same light beam travelling at the same speed? Einstein put forward two simple **postulates**. The two postulates of special relativity can be abbreviated to:

1. The laws of physics are the same in all inertial frames of reference.
2. The speed of light is the same to all observers.

Einstein realised that accepting both of these postulates implied that space and time were not absolute and independent, but were related in some way. For example, consider a train that has a flashing light bulb set right in the centre of the carriage. If Cindy, who is outside the train, were to measure the light travelling at the same speed in the forwards and backwards directions, she will find that the light will reach the back wall first. This is because that wall is moving towards the light, whereas the front wall is moving away from the light, and so the light will take longer to catch up to it. This is shown in Figure 7.13.

FIGURE 7.13 Space and time are not absolute. Cindy measures the light reaching the back wall of the train carriage before it reaches the front wall, as the train is travelling forwards. An observer on the train would see the light reach the front and back walls at the same time.

Two events that are simultaneous in one frame of reference are not necessarily simultaneous in another. This implies that time measured in different frames of reference might not be the same. This lead to the conclusion that both time and space are related in a four-dimensional universe called **spacetime**.

EVIDENCE OF SPECIAL RELATIVITY

In general terms, due to Einstein's second postulate, for the speed of light to remain the same for all observers then either time or space will need to change to compensate for this. For example, consider the situation in Figure 7.14 where Amaya and Binh are riding in a spaceship that can travel at speeds close to the speed of light. Clare is going to watch from a space station, which according to Clare is a stationary frame of reference.

Amaya and Binh have taken along a light clock, which (it is assumed) Clare can read, even from a large distance away. A light clock 'ticks' each time the light pulse reflects off the bottom mirror. Clare also has an identical light clock in her own space station, which she can compare to Amaya's light clock. Like any light clock, this clock is governed by a regular oscillation that defines a period of time. Clare measures that the light pulses travel a zigzag path between the mirrors.

FIGURE 7.14 Time dilation: light clock

In their own frame of reference, Amaya and Binh measure a unit of time equal to t_a. Clare, from her frame of reference, measures a different time, t_c. The proper time, t_0, is the time measured by an observer at the same point in an inertial frame of reference; in this example it would be Amaya's and Binh's time, t_a. Clare's measured time, t_c, is greater than the time that Amaya and Binh measure, t_a, for the same event. This means the pulses in a light clock in a moving frame of reference have to travel further when observed from a stationary frame ($t = \frac{d}{c}$). Because of the constancy of the speed of light, this effectively means that time appears to have slowed in a moving frame.

This is known as **time dilation**, according to the equation:

$$t = t_0\gamma$$

where:

t_0 is the proper time in the moving frame

t is the time observed from the stationary frame

γ is the **Lorentz factor**:

$$\gamma = \frac{1}{\sqrt{1-\frac{v^2}{c^2}}}$$

where:

v is the speed of the moving frame of reference (m s^{-1})

c is the speed of light ($3.0 \times 10^8\,\text{m s}^{-1}$).

The situation above was from Clare's point of view, not Amaya's and Binh's. According to Amaya and Binh, as they look out their window at Clare in her space station receding from them, they can consider that it is they who are at rest and it is Clare and her space station that are moving away at a velocity near the speed of light. This is what Galileo's principle of relativity and Einstein's first postulate are all about. Observers in relative motion both measure time slowing in the other frame of reference.

In the Earth's atmosphere, high-energy cosmic rays interact with the nuclei of oxygen atoms 15 km above the surface of the Earth to create a cascade of high velocity subatomic particles. One of these particles is a muon, which is unstable. Time dilation explains how muons can reach the Earth's surface after originating 15 km up in the upper atmosphere, when they should all decay within 7 km of their origin according to classical physics. Within the stationary observer's frame of reference, the muons' lifetime is dilated so that they reach the detectors on the Earth's surface before we would expect them to have decayed. From the muons' frame of reference, the distance travelled is actually shorter. This is known as **length contraction**.

The theory of special relativity states that time and space are related. Motion affects space in the direction of travel. A moving object will appear to travel a shorter distance, by the inverse of the Lorentz factor, γ. The proper length, l_0, is the length measured by an observer at rest with respect to the object being measured.

Einstein's length contraction equation is given by:

$$l = \frac{l_0}{\gamma}$$

where:

l_0 is the proper length in the stationary frame

l is the contracted length as seen in the moving frame

γ is the Lorentz factor.

MOMENTUM AND ENERGY

As discussed above, the Lorentz factor is used to calculate the degree of time dilation and length contraction of an object moving relative to an observer. The problem with this equation can be seen if you consider a spaceship that reaches the speed of light: the length of the spaceship would therefore shrink to zero, and time inside it would appear to stop altogether. Einstein took this to mean that it is not possible to reach the speed of light in any real spaceship. However, the difficulties with time and length for the spaceship were not the only reasons Einstein came to this conclusion. If a rocket ship like the one in Figure 7.14 can travel at $0.99c$, why can't it simply turn on its rocket motor and accelerate up to c, or more? Einstein showed that as the speed of a spaceship approaches c, its momentum increases, but this is not reflected in a corresponding increase in speed.

Relativistic momentum (momentum described only by the theory of relativity) includes the Lorentz factor, γ, and hence, as more impulse (change in momentum) is added, the mass seems to increase towards infinity as the speed gets closer, but never equal, to c.

GO TO ➤ Year 11 Module 2

The relativistic momentum equation is:

$$p_v = \gamma m_0 v = \gamma p_0$$

where:

p_0 is the momentum, $m_0 v$, as you would define it in classical mechanics (kg ms^{-1}), where m_0 is the proper mass

p_v is the relativistic momentum (kg m s^{-1}).

A term called relativistic mass, γm_0, may be used to indicate the mass of an object that is moving. As the Lorentz factor increases with the increase in the velocity, then the relativistic mass also increases. Consider the example with the rocket ship that is attempting to increase its velocity to the speed of light. With the increase in the relativistic mass of the rocket ship, it becomes harder for the force of the engines to cause a change in velocity (and therefore a change in momentum, or impulse). The closer the rocket ship approaches to c, the greater the amount of impulse that is required to accelerate the ship to the speed of light. This was the reason, Einstein concluded, why an object can never reach the speed of light.

As the momentum of an object increases, so does its kinetic energy. Einstein showed, however, that the classical expression for kinetic energy was not correct at high speeds. The kinetic energy is given by:

$$K = (\gamma - 1)mc^2$$

From this equation, and by considering conservation of energy, Einstein found that the total energy of an object is given by:

$$E_{\text{total}} = K + E_{\text{rest}} = \gamma mc^2$$

The rest energy, which is the energy associated with the rest mass of an object, is given by:

$$E_{\text{rest}} = mc^2$$

Mass and energy are seen as different forms of the same thing. This means that mass, m, can be converted into energy, and energy can be converted into mass.

Nuclear fission and fusion reactions result in a **mass defect** (change). It is this difference in mass that is converted to the energy released in nuclear reactions. This mass is related to the energy produced according to:

$$\Delta E = \Delta mc^2$$

Nuclear fusion is the combining of light nuclei to form heavier nuclei. Extremely high temperatures are required for fusion to occur. This is the process occurring in stars. Hydrogen nuclei fuse to form deuterium. Further fusions result in the formation of isotopes of helium.

When a particle interacts with its antiparticle (such as an electron and a positron), **annihilation** occurs. For electron-positron annihilation, this interaction produces gamma rays. It may seem that mass is not conserved in this event as, while the particle–antiparticle pair has mass, gamma rays (i.e. light) do not. However, using Einstein's equation, as the mass of an object is a measure of its energy content, the mass of the electron and positron is converted into energy.

ISBN 978 1 4886 1936 6

WORKSHEET 7.1

Knowledge review—light and waves

1 Rearrange these colours in order of increasing wavelength.

green, indigo, red, orange, violet, yellow, blue

2 Rearrange these radiation types in order of increasing frequency.

X-rays, microwaves, blue light, radio, infrared, cosmic rays, ultraviolet

3 A 2.2 cm pin is placed 20 cm in front of a concave mirror whose focal length is 8.0 cm. Find the nature, location and size of the image formed by the mirror. (Hint: Remember the lens/mirror equation: $\frac{1}{f}=\frac{1}{u}+\frac{1}{v}$.)

4 Find the frequency of EMR whose wavelength is 1.25×10^{-3} m. What type of radiation is it?

5 A raindrop hits a calm water surface and creates a series of ripples. Which of the following quantities changes as the ripples move out from the starting point?

A wavelength

B amplitude

C frequency

D velocity

6 A point source of light has an intensity of $5.6\,\text{W}\,\text{m}^{-2}$ at a distance of 3.0 m. At what distance would an intensity of $8.6\,\text{W}\,\text{m}^{-2}$ be measured?

7 In a ripple tank, a set of parallel waves 1.5 cm apart is travelling towards a shallower region at $7.5\,\text{cm}\,\text{s}^{-1}$. Each wavefront makes an angle of 20° with the straight edge of the deeper section. The wave separation becomes 1.3 cm once it passes into shallower waters.

a What is the frequency of the waves in the deeper section?

b What is the frequency in the shallower section?

c What angle do the wavefronts make with the boundary of the shallower section? (Hint: Use a variation of Snell's law: $n_1 \sin \theta_1 = n_2 \sin \theta_2$.)

8 Sort the following waves into transverse or longitudinal.

sound, radio, guitar-string vibrations, water, sonar, Mexican wave

9 After waiting for a while, a surfer sees a gentle swell coming in. As it approaches, the surfer first starts to move in which direction?

A backwards

B forwards

C up

D down

 ISBN 978 1 4886 1936 6

WORKSHEET 7.2

Investigating spectra

There are two types of spectra covered in this physics course: emission and absorption.

An emission spectrum can be created by passing an electric current through a gas in a discharge tube. A very large potential difference generally needs to be applied across the contacts at either end of the discharge tube to start the current.

1 The discharge inside the tube looks like a steady spark or 'plasma'. Describe what happens to the atoms or molecules in the gas when the discharge takes place.

2 Plasma is sometimes referred to as a 'fourth state of matter'. Explain what is meant by this.

3 The light coming from the tube is created when excited atoms transition to lower energy states. Each emitted wavelength corresponds to a specific transition. If a wavelength of 451 nm was found in an emission spectrum, what is the size of the energy transition (in joules) that created it?

4 A more common unit of energy employed in this area is the electron volt (eV). Why is this energy unit given this name? How many joules are equal to one electron volt?

5 Convert the energy found in question 3 into electron volts.

The wavelengths of light emitted from a discharge tube are generally measured with a spectroscope. Newton originally created a spectrum by passing white light through a glass prism. A diffraction grating, either etched on transparent plastic or scratched on glass, will also produce a spectrum.

6 Give two reasons why a diffraction grating produces a much better spectrum than a glass prism.

Your school may also have a HeNe laser, which emits light at 633 nm. If not, a red laser pointer in a PowerPoint presentation 'clicker' might be available. These generally emit light at around 650 nm. Shine the laser through a diffraction grating at a white screen about a metre away.

 Exercise care when using a laser. Do not shine it into anyone's eyes!

You should be able to see a bright spot directly in front of the laser. In addition, you should see a pair of red spots equally spaced to either side of the central spot. (You may need to rotate the diffraction grating to see them). There may even be another pair of fainter spots further out.

7 Measure the distance of a spot from the centre of the screen (x) and the distance of the screen from the grating (L). Use the equation $d \sin\theta = m\lambda$ to find d, the spacing between the lines on the diffraction grating. ($m = 1$ for the first spot.)

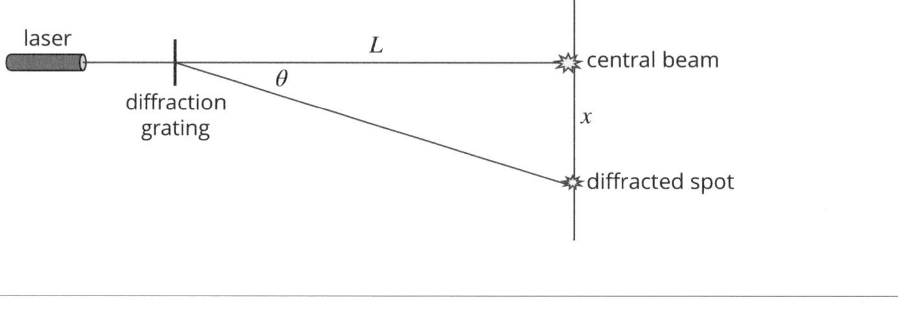

Now use your spectroscope to examine some different light sources. Discharge tubes would be ideal—as long as you can find a really dark room. Other sources to try would be:

- incandescent light globes
- fluorescent tubes
- low-energy light globes
- LED lights
- LED traffic lights
- street lights—especially the yellow ones
- sunlight reflected from white paper. Do not look at the sun!

8 Do you see distinct emission lines in all the spectra? Which ones do not have distinct lines?

An absorption spectrum is created when light of a range of wavelengths passes through a gas or plasma. The gas may absorb some specific wavelengths, leaving dark lines in the continuous spectrum. Each of these dark lines will correspond to an emission line belonging to the gas. When a specific wavelength is absorbed, the atom is excited into a higher energy state. It immediately returns to a lower energy level by emitting the energy just received—often producing light of the same wavelength.

 ISBN 978 1 4886 1936 6

9 If the same colour that was absorbed is re-radiated, why are dark lines formed at all? Why don't the re-emitted waves fill in the dark gaps?

Stars have distinct colours, each colour being the peak intensity of the star's spectrum. Wien's law can be used to calculate the surface temperature of a star. Wien's law is:

$$\lambda_{max} = \frac{b}{T}$$

where λ_{max} is the peak wavelength (m), $b = 2.9 \times 10^{-3}$ m K and T is the temperature (K).

10 The Sun's surface temperature is about 6000 K. What is its peak wavelength? What colour does this correspond to? Comment.

The absorption spectra of stars tell us exactly what elements are present in the gases of their outer stellar atmospheres. Their width or 'darkness' indicates how much of each element is present. In addition, the mix of absorption lines can tell us a good deal about the temperature and type of star. For example, a blue star with few lines other than helium will indicate a young star, whereas a yellow star with many lines of metals will be an older star, created from the leftover remnants of a supernova.

11 If the lines in an absorption spectrum are shifted from their usual position, it would indicate that a star is moving towards or away from the observer. What is the name of this effect? What would a shift to the blue end of the spectrum indicate?

RATING MY LEARNING	My understanding improved	Not confident ◄——► Very confident ○ ○ ○ ○ ○	I answered questions without help	Not confident ◄——► Very confident ○ ○ ○ ○ ○	I corrected my errors without help	Not confident ◄——► Very confident ○ ○ ○ ○ ○

WORKSHEET 7.3

Investigating diffraction

1 a Given the wavelength range of visible light, what size aperture or obstacle is needed to significantly diffract light?

b How does this affect optical microscopes?

2 For the obstacles shown below, draw the expected diffraction patterns.

(a) **(b)**

3 In the early days of radio, particularly in Europe, long-wave radio signals used to broadcast in the 140–253 kHz range. This frequency range is not used often these days due to the higher quality sounds in FM and the advent of digital, internet and satellite signals. Very high frequency (VHF) signals, with frequencies of 30–300 MHz, are often used for walkie–talkies and in the past were used for television.

a Calculate the wavelength range of long-wave and VHF signals.

b Explain why the long-wave radio signals do not need many repeater stations to travel a long distance, and why they experience less interference from obstacles in their path.

RATING MY LEARNING	My understanding improved	Not confident ◄——► Very confident ○ ○ ○ ○ ○	I answered questions without help	Not confident ◄——► Very confident ○ ○ ○ ○ ○	I corrected my errors without help	Not confident ◄——► Very confident ○ ○ ○ ○ ○

 ISBN 978 1 4886 1936 6

WORKSHEET 7.4

Evidence for the wave model of light

The following is a list of some of the behaviours or properties of light you may have examined.

straight-line propagation
reflection
Compton effect
absorption
double refraction and polarisation
diffraction
colour
dispersion
refraction
inverse square law
photoelectric effect
Wien's law
two-slit interference
thin-film interference
propagation through a vacuum

a Find out which of these behaviours or properties would have been known to Isaac Newton and/or Christiaan Huygens; which would have been known since ancient times; and which would not have been known until more recent times. If possible, find a date by which the phenomenon was first noted or discovered.

Known since ancient times	Known to Newton and Huygens	Unknown to Newton and Huygens

b Briefly outline the model for light formulated by Newton and Huygens.

c Based on your learning from parts a and b sort the behaviours of light listed above into three categories by writing the appropriate letter by each behaviour:

supports Newton's model (N)

supports Huygens' model (H)

could support either model (E).

RATING MY LEARNING	My understanding improved	Not confident ⟷ Very confident ○ ○ ○ ○ ○	I answered questions without help	Not confident ⟷ Very confident ○ ○ ○ ○ ○	I corrected my errors without help	Not confident ⟷ Very confident ○ ○ ○ ○ ○

WORKSHEET 7.5

Diffraction and interference

DIFFRACTION GRATINGS

Diffraction gratings work on the same principle as the double-slit experiment, except that they have many lines or slits. A diffraction grating is defined by its number of lines per cm.

Consider a grating with 4000 lines per cm. Then the slit spacing is given as:

$$d = \frac{1}{4000} = 2.5 \times 10^{-4}\,\text{cm} = 2.5\,\mu\text{m}$$

The more slits a diffraction grating has, the narrower the interference fringe and the more precise optical spectroscopy can be. A simple arrangement, where the detector angle can be changed to pick up specific wavelengths, is shown below.

The collimator is a lens arrangement which collects the light using one lens and focuses the light onto the diffraction grating using a second lens.

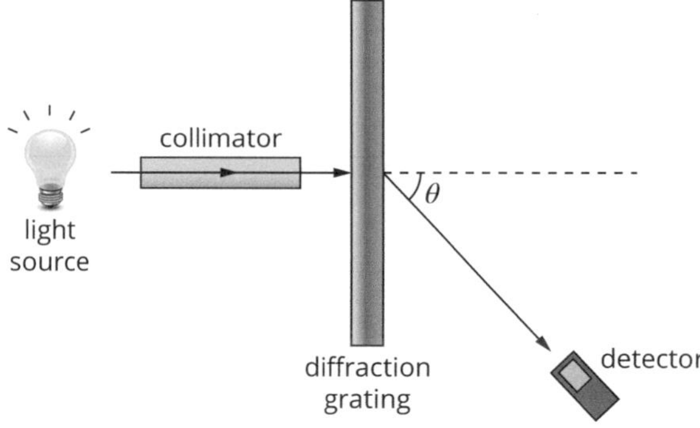

As for a double slit, the condition for constructive interference in a diffraction grating is given by $m\lambda = d\sin\theta$.

1 A HeNe laser of wavelength 633.2 nm is incident on a diffraction grating with 5000 lines cm^{-1}. At what angle does the detector need to be to observe the first-order line of this wavelength?

2 A white light source is shone through the diffraction grating. What colour will be seen at the angle calculated in question 1?

3 **a** The angle of the detector is changed to 40°. What first-order wavelength will be observed?

b What second-order wavelength could be detected?

ISBN 978 1 4886 1936 6

4 In X-ray diffraction, the spacing between atoms becomes the diffraction grating. A crystal is rotated through an angle θ, and the angles at which constructive interference occur give information about the atomic spacing and the crystal structure.

a In the diagram below, look at the path difference between the two rays and write the condition for constructive interference in this case.

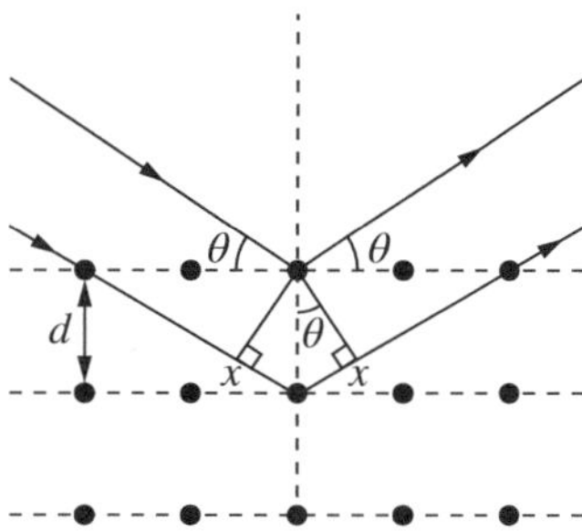

b An NaCl crystal has a spacing between adjacent atoms of 0.2814 nm. What wavelength will the X-ray need to be to observe constructive interference at an angle of 40°? Assume first-order effects only.

INTERFERENCE OF LIGHT

5 What property of EMR does Young's double-slit experiment show?

6 For the double slit and screen shown, draw the wavefronts and indicate the regions of constructive and destructive interference.

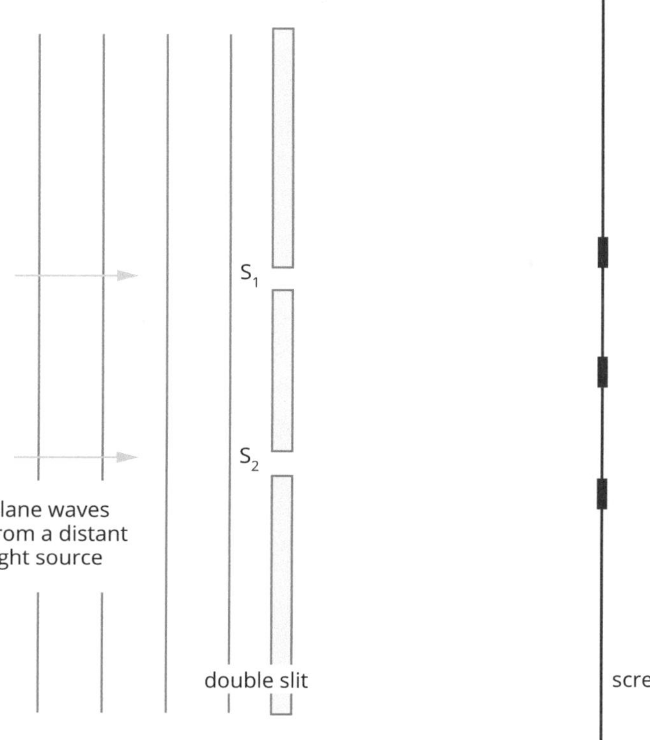

7 The direction of the wave can be indicated by a ray. Consider two waves, incident on slits S_1 and S_2 at an angle, as shown in the following diagram.

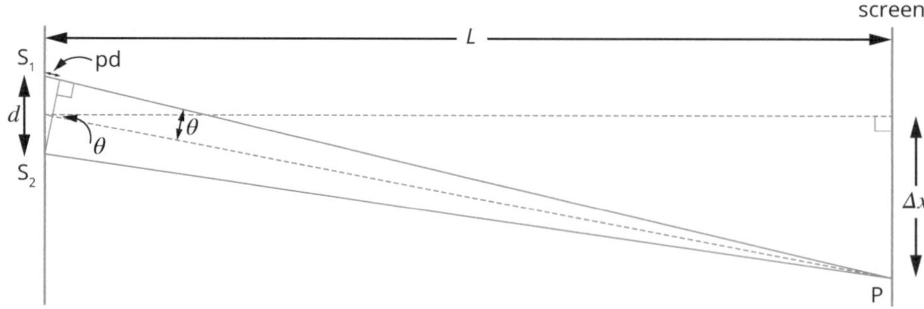

a What does the path difference have to be for constructive interference? Explain your answer.

b What does the path difference have to be for destructive interference? Explain your answer.

8 An interference pattern from a red laser is given below.

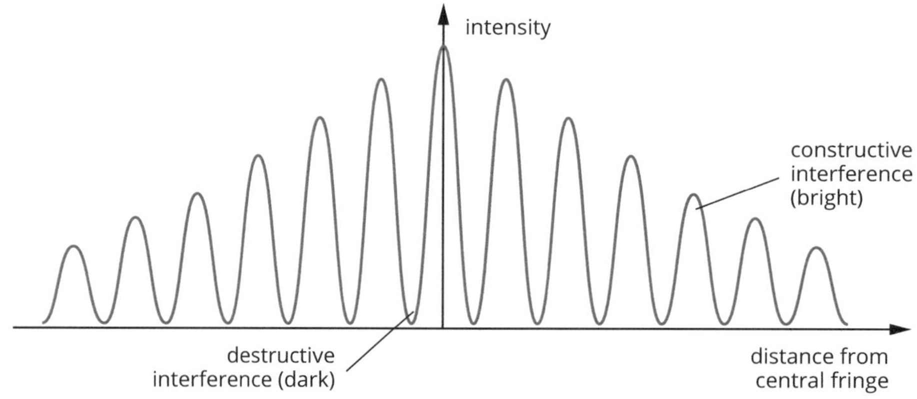

How would you expect the pattern to change if:

a a green laser is used instead of a red laser

b the red laser is used but the slit spacing is decreased

c the red laser is used and the slits are moved closer to the screen.

9 A blue semiconductor laser of 450 nm is directed through a thin pair of slits, 70 μm apart, onto a screen that is 2.5 m away. Calculate the fringe spacing.

 ISBN 978 1 4886 1936 6

10 Light of an unknown wavelength is emitted through a pair of thin slits 30 μm apart. The screen is 1.5 m away and the fringe separation is 2.0 cm. Determine the wavelength and colour of the laser.

____

____

____

11 A Young's double-slit experiment was performed with a semiconductor laser diode of unknown wavelength. To reduce experimental error, the student decided to take a series of measurements. The double slits were 50 μm apart. The student changed the distance of the screen from the slits and measured the fringe separation. Plot the data on a suitable graph and determine the wavelength using a line of best fit and its gradient.

***L* (m)**	**Δ*x* (cm)**
1.0	1.3
1.5	2.0
2.0	2.6
2.5	3.3
3.0	3.9

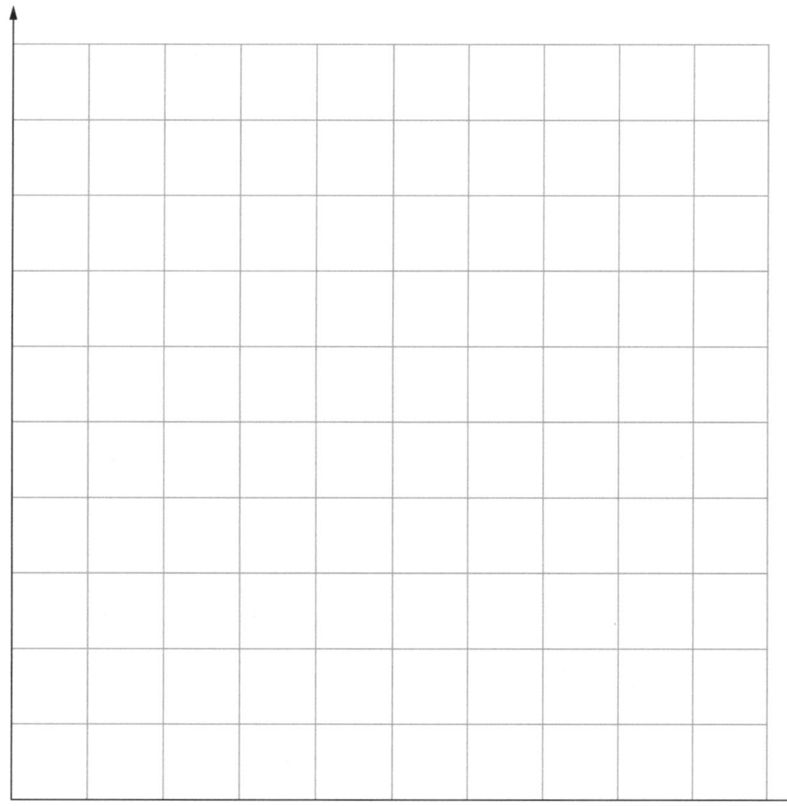

RATING MY LEARNING	My understanding improved	Not confident ◄——► Very confident ○ ○ ○ ○ ○	I answered questions without help	Not confident ◄——► Very confident ○ ○ ○ ○ ○	I corrected my errors without help	Not confident ◄——► Very confident ○ ○ ○ ○ ○

ISBN 978 1 4886 1936 6

WORKSHEET 7.6

Black-body radiation

Energy density versus wavelength curves are shown for black-body radiation at various temperatures.

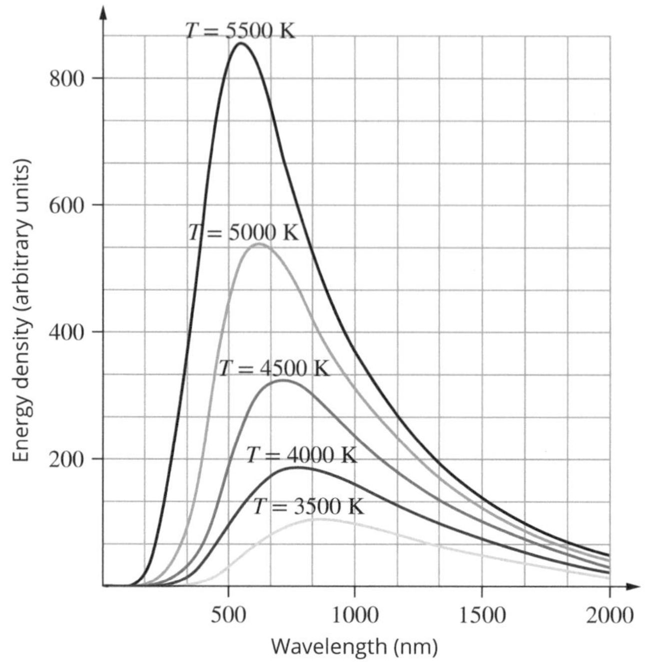

1 Which (one or more) of the following statements about a black body are true?

A The shape of an emission spectrum depends only on temperature.

B Black-body radiation depends on the type of material and its shape.

C A black body only absorbs EMR.

D A black body is completely black.

E A black body emits and absorbs EMR.

F A black body emits EMR of different wavelengths.

2 Using Wien's displacement law, calculate the expected peak wavelength for:

a 3500 K

____

b 4500 K

____

c 5500 K.

____

3 In classical physics, thermal radiation is emitted by accelerating charges near the surface of an object. The radiated energy from a black body can be approximated by standing waves or resonant modes within a cavity. At long wavelengths there is a reasonable agreement between experimental data and classical theory.

Explain why the classical theory does not agree at shorter wavelengths.

____

____

____

____

 ISBN 978 1 4886 1936 6

4 a State Planck's two fundamental postulates that are used to explain the shape of the black-body curve.

b Explain how this solved the issue at short wavelengths.

5 a A quantum of light has an energy of 2.8×10^{-19} J. Convert this to electron-volts.

b Calculate the energy of a photon of green light, wavelength 510 nm.

6 Our Sun has a surface temperature of 5800 K. Assuming the Sun behaves as a black body, calculate its peak wavelength and hence determine the energy of an individual photon from the Sun in eV.

7 An old-fashioned tungsten filament globe produces light by heating a wire. It gives off a similar emission spectrum to the Sun, hence its peak wavelength is similar. Assume for the purposes of this calculation it is the same. Determine how many photons are emitted from a 60 W lamp in 1 hour.

8 At room temperature, most objects emit in the infrared spectrum (hence the use of infrared goggles for 'seeing' in the dark). Assuming these objects can be approximated by a black body and that room temperature is 20°C, calculate the peak wavelength and energy of an emitted photon in J and eV.

RATING MY LEARNING	My understanding improved	Not confident ←→ Very confident ○ ○ ○ ○ ○	I answered questions without help	Not confident ←→ Very confident ○ ○ ○ ○ ○	I corrected my errors without help	Not confident ←→ Very confident ○ ○ ○ ○ ○

WORKSHEET 7.7

The photoelectric effect

A circuit for measuring the photoelectric effect is shown in the figure below.

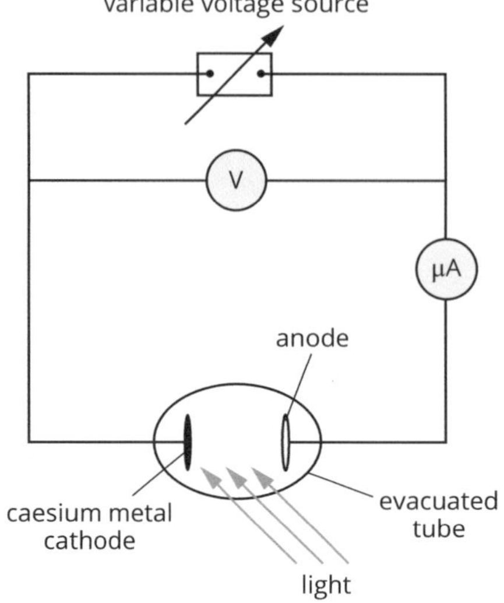

1 The following are some observations of the photoelectric effect. What suitable conclusions could be drawn from these observations?

a There is a threshold frequency below which no photoelectrons are emitted, even if the intensity is increased.

b Above the threshold frequency, increasing the intensity increases the number of photoelectrons produced.

c When a reverse bias is applied, the cathode has a positive potential. There is still a photocurrent at a small reverse bias.

2 If the stopping voltage is 1.8 V, calculate the maximum kinetic energy of the photoelectrons in eV and joules.

3 Two photoelectric effect experiments are conducted where the number of photons incident on the metal is the same. In the first, a light of frequency f_1 is used, and a photocurrent I_1 and stopping potential V_1 are measured. In the second, a lower frequency f_2 is used, and a photocurrent I_2 and stopping potential V_2 are measured. Which is the most correct statement? Explain your answer.

A photocurrent $I_2 < I_1$ and stopping potential $V_2 < V_1$

B photocurrent $I_2 < I_1$ and stopping potential $V_2 = V_1$

C photocurrent $I_2 = I_1$ and stopping potential $V_2 < V_1$

D photocurrent $I_2 = I_1$ and stopping potential $V_2 > V_1$

 ISBN 978 1 4886 1936 6

4 Explain why the wave model cannot be used to explain the photoelectric effect.

5 The work function for the following metals is given below:

calcium	2.9 eV	niobium	4.3 eV
copper	4.7 eV	gold	5.1 eV

a Which metal will emit photoelectrons if a light of 1.09×10^{15} Hz is incident on it?

b For the metals that will eject photoelectrons, what will be the maximum kinetic energy in eV?

6 Beryllium has a threshold frequency of 1.20×10^{15} Hz. Calculate the maximum kinetic energy (eV) of the ejected photoelectrons when UV light of wavelength 200 nm is incident on the beryllium cathode.

7 Determine the work functions of potassium and magnesium from the graph at right, using two methods.

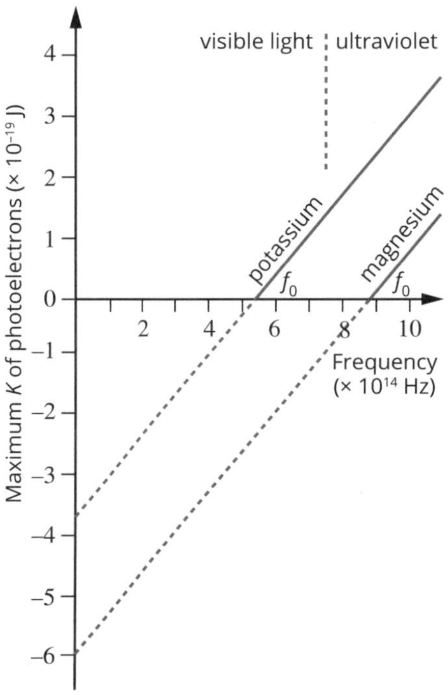

RATING MY LEARNING	My understanding improved	Not confident ◄—► Very confident ○ ○ ○ ○ ○	I answered questions without help	Not confident ◄—► Very confident ○ ○ ○ ○ ○	I corrected my errors without help	Not confident ◄—► Very confident ○ ○ ○ ○ ○

WORKSHEET 7.8

It's all relative—special relativity

If you watch the activity inside a moving train from the platform outside, you would generally think that you would need to add the velocity of the train to any velocity measured inside the train in order to find the final velocity based on your stationary frame of reference. Einstein realised that this idea of adding velocities was based on a questionable assumption: that time and space are 'absolute' and 'uniform'. Newton also realised that his work was based on this assumption, but he felt, like most of us, that it was a reasonable one. Certainly, at the everyday speeds of a train and moving passengers, it is.

However, Einstein postulated that light will always travel at the same velocity, no matter what frame of reference we measure it in, and experiments since have established that this is the case. This has strange implications for the nature of space and time. In particular, they become 'relative'.

In this simulation, the speed of light is reduced to 'ordinary' speeds. Your simulation is based on Einstein's own discussions of a light flashing in a train. The light, in the centre of the train carriage, is measured by observers inside the train to reach the ends of the carriage at the same time. But what do observers, such as Chloe who is outside this moving train, measure?

Observers only 'see' the light when it reaches their eyes. However, they can calculate from the known distance and speed of light how long ago the light actually reached an object. This is referred to as the look-back time.

In this simulation, you will calculate the time at which the two light flashes reach the front and rear walls of the carriage and compare these values for each observer.

In order to calculate the required times, you can make use of a convenient simplification. While Anna and Ben, who are travelling on the train, and Chloe all measure the speed of light as c, the time taken for the light to reach the end of the carriage, as Chloe measures it, can be found from the apparent relative speed of light to the carriage. Chloe measures the light moving backwards at c and the train moving forwards at v, and so the time for the light to reach the rear of the carriage is obtained by dividing the relevant distance by $c + v$ (or by $c - v$ for the light flash that is moving forwards).

No one actually sees the light travelling at $c + v$. Chloe measures the train moving forwards and the light moving backwards so that, to her, the relative speed of light and train appears to be $c + v$. Chloe also measures the train contracted by the Lorentz factor γ, and so the length to be divided by $c + v$ is $\frac{l}{\gamma}$.

1 For this simulation, rather than having impossibly fast trains, light will be slowed down to more 'ordinary' speeds. Is this a reasonable constraint?

 ISBN 978 1 4886 1936 6

2 Complete the following table by calculating the relevant quantities for train speeds between $0\,\mathrm{m\,s^{-1}}$ and $100\,\mathrm{m\,s^{-1}}$ for a speed of light, c, equal to $100\,\mathrm{m\,s^{-1}}$. The quantities are as follows:

T_A—the time for the light to reach the ends of the carriage as measured by Anna and Ben

T_{CR}—the time for the light to reach the rear wall as measured by Chloe

T_{CF}—the time for the light to reach the front wall as measured by Chloe

T—the difference in times as measured by Chloe.

l (m)	c ($\mathrm{m\,s^{-1}}$)	v ($\mathrm{m\,s^{-1}}$)	γ	$T_A = \frac{l}{c}$ (s)	$T_{CR} = \frac{\frac{l}{\gamma}}{(c+v)}$ (s)	$T_{CF} = \frac{\frac{l}{\gamma}}{(c-v)}$ (s)	$T = T_{CF} - T_{CR}$ (s)
10	100	0					
10	100	10					
10	100	20					
10	100	30					
10	100	40					
10	100	50					
10	100	60					
10	100	70					
10	100	80					
10	100	90					
10	100	99					

3 Plot graphs of T_A, T_{CR}, T_{CF} and T against the speed of the train on the graph provided. Label the axes appropriately with both quantities and units.

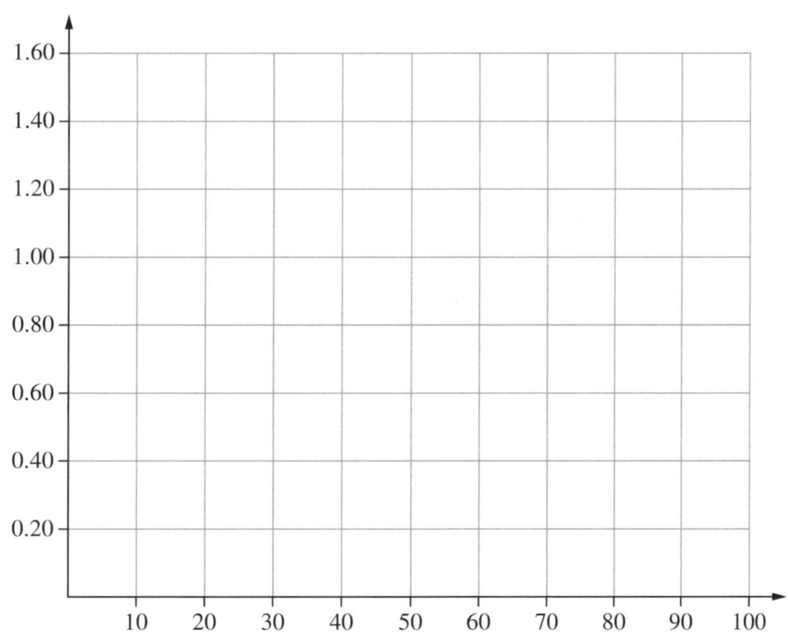

ISBN 978 1 4886 1936 6

4 What happens when the speed of the train is equal to the speed of light? Why is this? What happens, and why, when the train has a speed faster than light?

5 For the situation when the train is at rest, do the observers inside and outside the train agree on their calculations of when the light hit the front and rear walls?

6 Describe the situation as observed by the different observers when the train is travelling slowly. How do Chloe's observations compare with Anna's and Ben's? What happens as you increase the speed of the train?

7 Is the non-simultaneity that Chloe observes just a 'trick' of look-back effects? Explain your answer.

8 What happens in your simulation if you put in the accepted value for the speed of light and keep the speed of the train at the same magnitude as that used in the table?

9 Investigate and describe one type of experimental evidence for time dilation and/or length contraction. Examples include:

Hafele–Keating experiment

Frisch–Smith experiment.

RATING MY LEARNING	My understanding improved	Not confident ◄──► Very confident ○ ○ ○ ○ ○	I answered questions without help	Not confident ◄──► Very confident ○ ○ ○ ○ ○	I corrected my errors without help	Not confident ◄──► Very confident ○ ○ ○ ○ ○

 ISBN 978 1 4886 1936 6

WORKSHEET 7.9

Literacy review—the language of light and matter

Complete the table with a definition of each of the included terms.

absorption spectrum	
aether	
coherent	
corpuscles	
diffraction	
electromagnetic radiation	
emission spectra	
frame of reference	
in phase	
interference	
interferometer	
length contraction	
photoelectric effect	
photon	
proper time	
relativistic mass	
simultaneity	
spacetime	
threshold frequency	
time dilation	

RATING MY LEARNING	My understanding improved	Not confident ◄——► Very confident ○ ○ ○ ○ ○	I answered questions without help	Not confident ◄——► Very confident ○ ○ ○ ○ ○	I corrected my errors without help	Not confident ◄——► Very confident ○ ○ ○ ○ ○

ISBN 978 1 4886 1936 6

WORKSHEET 7.10

Thinking about my learning

On completion of Module 7: The nature of light, you should be able to describe, explain and apply the relevant scientific ideas. You should also be able to interpret, analyse and evaluate data.

1 The table lists the key knowledge covered in this module. Read each and reflect on how well you understand each concept. Rate your learning by shading the circle that corresponds to your level of understanding for each concept. It may be helpful to use colour as a visual representation. For example:

- green—very confident
- orange—in the middle
- red—starting to develop.

Concept focus	Rate my learning				
	starting to develop ◄———► very confident				
the electromagnetic spectrum	○	○	○	○	○
spectroscopy	○	○	○	○	○
use the spectra of stars to evaluate their surface temperature, velocity, density and composition	○	○	○	○	○
diffraction of light and Young's double slit experiment	○	○	○	○	○
polarisation	○	○	○	○	○
black-body radiation	○	○	○	○	○
photoelectric effect	○	○	○	○	○
the differences between the wave and particle models of light, and the development of the quantum model	○	○	○	○	○
Einstein's postulates on special relativity	○	○	○	○	○
time dilation	○	○	○	○	○
length contraction	○	○	○	○	○
relativistic momentum	○	○	○	○	○
Einstein's mass energy equivalence, $E = mc^2$	○	○	○	○	○

2 Consider points you have shaded from starting to develop to middle-level understanding. List specific ideas that you found challenging.

3 Write down two different strategies that you will apply to help further your understanding of these ideas.

ISBN 978 1 4886 1936 6

PRACTICAL ACTIVITY 7.1

Measuring the speed of light

Suggested duration: 25 minutes

INTRODUCTION

Today, manufacturers of high-quality science equipment produce apparatus for the laboratory measurement of the speed of light. These rely on the high-speed measurement of the reflection of monochromatic light over a measured distance or Foucault's method of rotating and fixed mirrors.

There is a simpler method that, while making some assumptions, can determine the speed of light with reasonable precision based on light's wave-like behaviour.

MATERIALS

- microwave oven—must be of the type that requires a turntable
- packet or two of marshmallows—you can also use a large chocolate block or choc chips
- microwave-safe plate or dish
- ruler
- heat-proof or oven-proof gloves

PURPOSE

To determine the speed of light based on light's wave-like behaviour.

Since microwaves are a form of EMR, as is visible light, determining the speed of microwave radiation can be used to determine the speed of light.

PROCEDURE

1 Remove the turntable from the microwave.

2 Open the bag of marshmallows and arrange them in the base of the dish, completely covering it with one layer of the marshmallows. If you're using chocolate blocks, break the block up and arrange the individual squares in a similar way.

3 Place the dish in the microwave and cook on a low power level. Microwaves don't heat evenly and normally require something like a turntable to ensure that the heat is distributed evenly. As the dish isn't being rotated, the marshmallows will begin to melt unevenly, melting more quickly in the 'hottest' parts of the oven. Continue to heat the marshmallows or chocolate until around four or five distinct melted spots are observed.

4 Carefully remove the dish from the microwave using the oven-proof gloves to avoid being burnt. Allow the dish to cool.

5 Using the ruler, measure the distance between two nearby melted spots and record them in Table 1 in the Data and analysis section. Repeat this for the distance between two different melted spots, until you have filled the table.

6 Record an estimate of the error in each of your distance measurements and calculate the percentage uncertainty in each measurement. Record your results in Table 1.

7 Recall that the velocity of a wave is given by $v = f\lambda$, or for the speed of light, $c = f\lambda$. The frequency of the microwaves in the oven will be recorded on the specifications panel fixed to the back or side of the microwave oven. Record this in Table 1.

PRACTICAL ACTIVITY 7.1

DATA AND ANALYSIS

TABLE 1 Results and calculations

Distance between melted spots (cm)	Measurement uncertainty (cm)	% uncertainty	Distance between melted spots × 2 (m)	Microwave frequency (Hz)	Speed of light, c (m s^{-1})

1 The distance between melted spots corresponds to half the wavelength of the microwaves. Why is this the case? Use a diagram to support your explanation.

2 Using twice the distance between melted spots as the wavelength of the microwaves, in metres, calculate the speed of light for each pair of melted spots. Record the average value with an estimate of the uncertainty in the result.

CONCLUSION

1 How does your calculation of the speed of light compare with the currently accepted value for the speed of light in air?

2 Comment on the reliability of conclusions you can draw from this investigation.

CLEAN UP

Make sure to eat the marshmallows (or chocolate)!

RATING MY LEARNING	My understanding improved	Not confident ←→ Very confident ○ ○ ○ ○ ○	I answered questions without help	Not confident ←→ Very confident ○ ○ ○ ○ ○	I corrected my errors without help	Not confident ←→ Very confident ○ ○ ○ ○ ○

 ISBN 978 1 4886 1936 6

PRACTICAL ACTIVITY 7.2

Light and spectra

Suggested duration: 40 minutes

INTRODUCTION

In the first part of this activity, you will create and examine a continuous spectrum of light. In the second part, you will observe the spectral lines from gas emission tubes using a spectrometer.

PURPOSE

To investigate how emitted light can be used to identify an element and provide evidence for atomic energy levels.

Safety warning: handle hot spectral tubes with care. Do not touch any tube with bare hands—use a soft cloth to handle the tubes or they may be permanently damaged. Read all accompanying safety notes before starting the experiment.

MATERIALS

- clear filament globe and power supply (a standard light box will also work)
- screen with single slit
- solid screen
- convex lens
- prism
- spectral tubes
- spectral tube power supply and mount
- spectroscope, diffraction grating glasses; or spectrometer and computing device
- multiclamp
- retort stand

Do not exceed the rated power of the globe supplied.

PROCEDURE

Part A: Creating a continuous spectrum

1 Set up the equipment as shown in the figure below. The filament globe acts as a white light source. After first passing through the slit, the light is dispersed into a continuous spectrum by the prism.

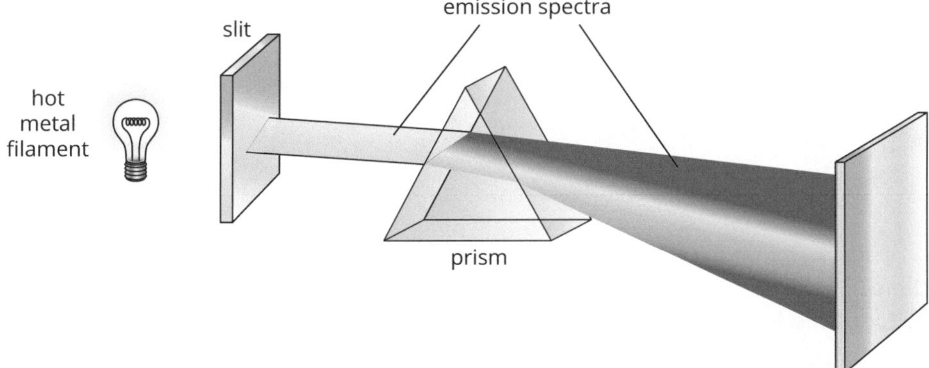

2 Place the convex lens between the slit and the prism. Move the lens backwards and forwards until a clear spectrum is produced on the solid screen. You may need to darken the room to see the spectrum clearly. In the Data and analysis section, record the order of the colours in the spectrum.

Part B: Observing spectral lines from gas emission tubes

Set up the spectral tube power supply and mount with a spectral tube—your teacher may have already done this. Do not turn on the tubes at this stage. The spectral tubes should only be illuminated for short times when they are about to be observed.

If you have a hand-held spectroscope or diffraction glasses, no further preparation is needed.

PRACTICAL ACTIVITY 7.2

Using a spectrometer

1 Turn on the spectrometer and connect it to your computing device.

2 Connect the fibre-optic cable to the spectrometer following the manufacturer's instructions.

3 Use a multi clamp to secure the probe or rounded end of the fibre-optic cable on a retort stand, as shown in the figure below.

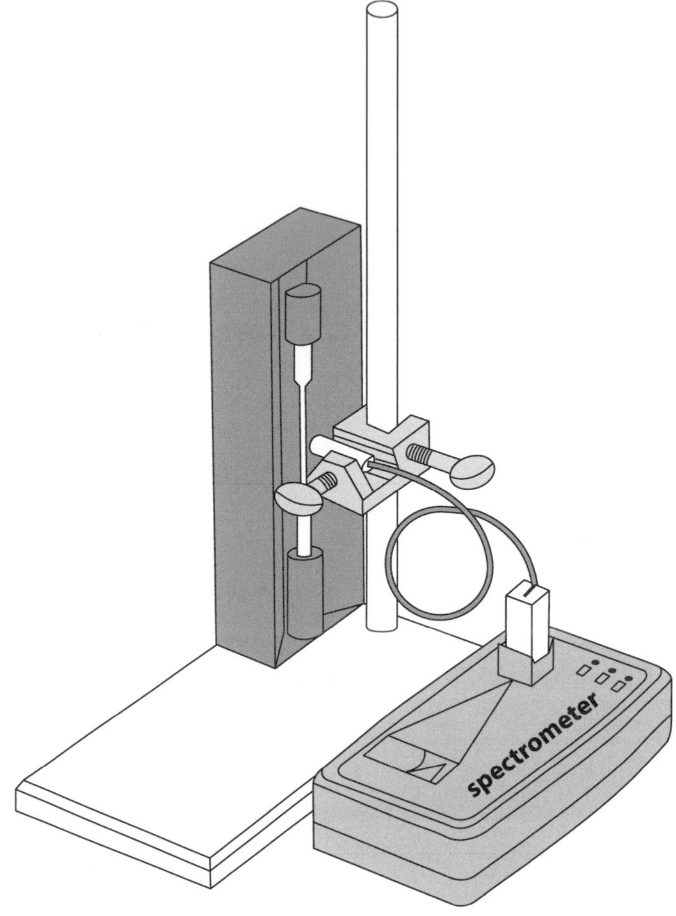

4 Place the end of the probe a distance of 2 cm or less from the gas tube. Adjust the probe to point towards the tube. Do not allow the probe to directly touch the tube.

5 Open the spectrometry application and choose 'analyse light', following your manufacturer's specific instructions.

6 Turn on the power to the spectral tube mount (your teacher may wish to do this—always check!). Start recording with the spectrometer software or start carefully observing the spectrum produced by the light.

7 Adjust the fibre-optic cable or angle of observation until a clear spectrum is seen.

DATA AND ANALYSIS

Part A: Creating a continuous spectrum

1 What is the order of the colours produced?

2 What does the order of the colours imply in terms of the respective wavelengths of the light?

3 What effect would placing a cloud of gas between the source and the slits have on the spectrum?

 ISBN 978 1 4886 1936 6

4 If possible, direct a beam of sunlight through the slits so a spectrum is formed on the viewing screen. Is this spectrum the same as the one from the filament globe? Comment on any differences.

Be very careful not to look directly at the sunlight.

Part B: Observing spectral lines from gas emission tubes

1 Record up to five significant peaks or lines in Table 1. Using the tools available to you, measure and record the corresponding wavelength for each line.

TABLE 1 Emission spectra

Colour of peak or line	Wavelength (nm)	Energy (J)

2 Based on your observations, what gas is contained in the spectral tube? You may need to check a suitable reference to find the expected spectrum for particular gases. (Some spectrometer software will include profiles for standard gases.)

3 Astronomers use high-powered spectrometers to analyse light throughout space. Explain how it is possible for an element to have the same line emission pattern every time it is energised, whether the element is in outer space or in a gas spectrum tube in a classroom.

4 For each spectral line identified in Table 1, calculate the equivalent energy of the emitted photons.

CONCLUSION

1 Using your particular results and observations, explain how spectroscopy can be used to identify particular elements.

2 What types of spectra are those produced by the spectral tubes and by sunlight: continuous, emission or absorption?

RATING MY LEARNING	My understanding improved	Not confident ◄—► Very confident ○ ○ ○ ○ ○	I answered questions without help	Not confident ◄—► Very confident ○ ○ ○ ○ ○	I corrected my errors without help	Not confident ◄—► Very confident ○ ○ ○ ○ ○

PRACTICAL ACTIVITY 7.3

The diffraction of light

Suggested duration: 20 minutes

INTRODUCTION

Convincing evidence of the wave-like behaviour of light includes diffraction. Any wave form will diffract around the edges of a gap or obstacle forming interference patterns on the other side. Complete diffraction occurs when the width of the gap is less than the wavelength of the incident wave. Partial diffraction will occur when the gap is larger than the wavelength. In this activity you will undertake a qualitative investigation of diffraction.

MATERIALS

- lamp or laser
- two cards with thin, vertical slits (commercial types can be used)
- translucent screen
- coloured filters

PURPOSE

To investigate qualitatively the diffraction of light through a gap as a wave-like behaviour of light.

PROCEDURE

The room should be darkened for this activity.

1 Place a piece of thick card, A, with a vertical slit in front of a lamp. This card creates a narrow beam. You can use a laser instead of the lamp and card if one is available.

2 In front of card A, place another card with a narrower slit, B. The slit should be carefully cut with clear, sharp edges or sourced from a commercial supplier.

3 Stand a translucent white screen (a piece of tracing paper would do) about 1 m away from slit B. Turn on the lamp and observe the pattern on the screen.

4 Place coloured filters in front of the lamp and observe the effect on the patterns of light on the screen.

Warning: If using a laser as the light source, take particular care not to look into its beam or shine it directly into other people's eyes. Permanent damage to the eye could result.

DATA AND ANALYSIS

In the space below, describe and draw the patterns observed, paying particular attention to the relative spacing for each colour and the intensity of each bright spot.

If the room cannot be sufficiently darkened, repeat the experiment without the screen. Hold slit B close to the eye and look at the filament of the lamp through the single slit. Record your observations. Do not attempt this if using a laser!

ISBN 978 1 4886 1936 6

PRACTICAL ACTIVITY 7.3

CONCLUSION

1 Comment on the patterns observed. Do they confirm that diffraction occurs for light? Does this confirm or contradict the wave model of light?

2 Comment on the effect that the different colour filters has on the spacing of the pattern.

3 Would there be any change in the pattern if the distance between card A and card B (or between the laser and card B) were changed? Try it if time is available.

4 Would there be any change in the pattern if the distance between card B and the viewing screen were changed? Try it if time is available.

RATING MY LEARNING	My understanding improved	Not confident ◄──► Very confident ○ ○ ○ ○ ○	I answered questions without help	Not confident ◄──► Very confident ○ ○ ○ ○ ○	I corrected my errors without help	Not confident ◄──► Very confident ○ ○ ○ ○ ○

PRACTICAL ACTIVITY 7.4

Interference of light—Young's double-slit experiment

Suggested duration: 45 minutes (one set of slits)

INTRODUCTION

In this activity, a key wave behaviour of light will be investigated. In 1801, Thomas Young obtained convincing evidence of the wave nature of light. In his experiment, light from a single source fell on a slide containing two closely spaced slits. If light consists of tiny particles (or 'corpuscles', as described by Isaac Newton), we might expect to see two bright lines on a screen placed behind the slits. Young observed a series of bright lines. He was able to explain this result as a wave interference phenomenon. Because of diffraction, the waves leaving the two small slits spread out from the edges of the slits. This is equivalent to the interference pattern of ripples produced when two rocks are thrown into a pond.

PURPOSE

To investigate qualitatively the interference of light through two parallel slits as a wave-like behaviour of light.

BACKGROUND

The equation that describes the conditions for constructive interference (interference maxima) in a double-slit interference pattern is:

$d \sin\theta = m\lambda$

where:

d is the spacing between the slits

θ is the angle at the centre line and the mth-order maximum in the interference pattern

m is an integer ($m = 0, 1, 2...$)

λ is the wavelength of the coherent light source.

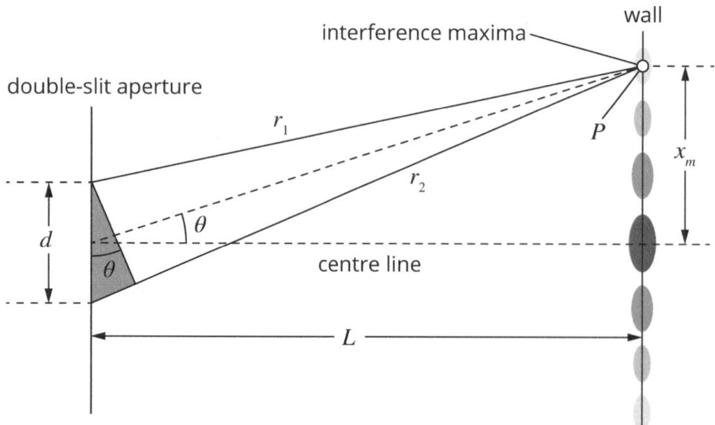

In practice, the angular position θ of each interference maximum can be expressed in terms of the linear position x, where x_m is the linear distance between the zeroth-order maxima and the mth-order maximum observed in the interference pattern on a screen some distance L from the double-slit aperture and:

$$x_m = m\frac{\lambda L}{d}$$

MATERIALS

- double slits—preformed, of various sizes
- retort stands
- three-finger clamps
- laser pointer/s with known wavelength (a spectrometer can be used to confirm the wavelength)
- white paper
- pencil
- ruler
- tape
- measuring tape

Warning: If using a laser as the light source, take particular care not to look into its beam or shine it directly into other people's eyes. Permanent damage to the eye could result.

Best results are obtained in a darkened room.

ISBN 978 1 4886 1936 6

PROCEDURE

1 Assemble your equipment and align the laser as shown in the figure below.

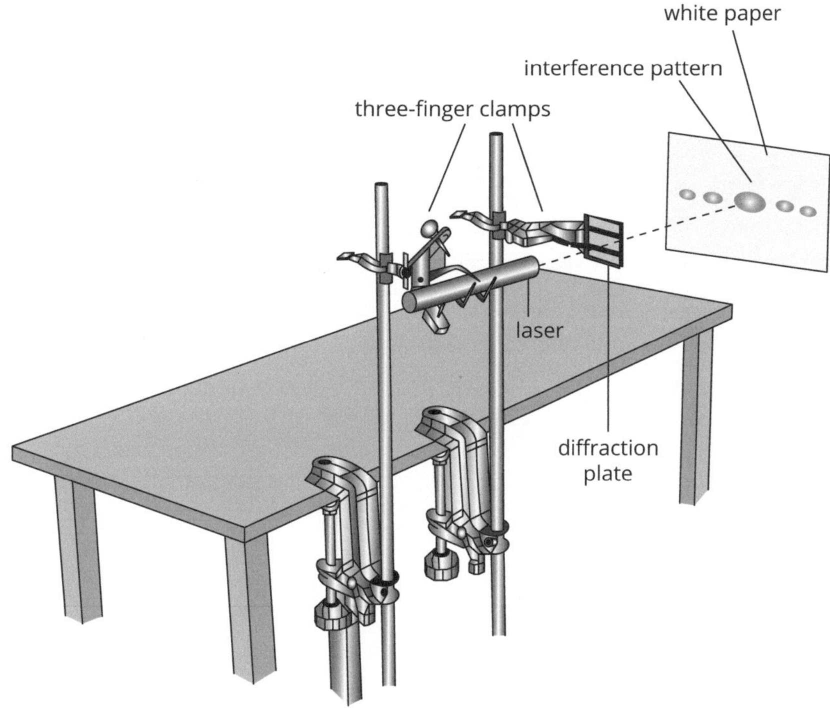

a Find an area on your lab table with enough space to align the components, with the laser aimed away from other lab groups and towards a wall or other flat rigid surface on which you can attach the white paper.

b Use tape to hold the white paper in place against the wall or other surface.

c Lightly clamp the edge of the interference plate (the edge parallel to the slits on the plate) in the three-finger clamp. Be certain the double-slit pattern on the plate is vertical and the fingers of the clamp touch only the edge of the plate. Over tightening may damage or distort the plate so be careful.

d Adjust the set-up so that the laser beam is perpendicular to the interference plate, and the white paper and interference plate are parallel to each other.

e Have at least 1m of space between the interference plate and the paper. Use as large a distance as possible while ensuring visibility of the light on the paper.

2 Adjust the laser so the beam shines on the interference slits. A clear image of the interference pattern should appear on the white paper. Record the distance between slits on the interference plate in the Data and analysis section.

> **Aim the laser slightly downwards at the diffraction plate to prevent stray reflection of the laser beam travelling upwards into a classmate's eyes. You might also want to place an object behind your set-up (behind the laser) to catch any reflection of the laser beam from the interference plate.**

3 Once aligned, tighten the set-up components so the laser and interference plate do not move during data collection.

4 Use the ruler and pencil to draw a horizontal line through the centre of the interference pattern on the paper.

5 Make a small mark on the line you just drew at the centre of the zeroth-order (central brightest) maximum.

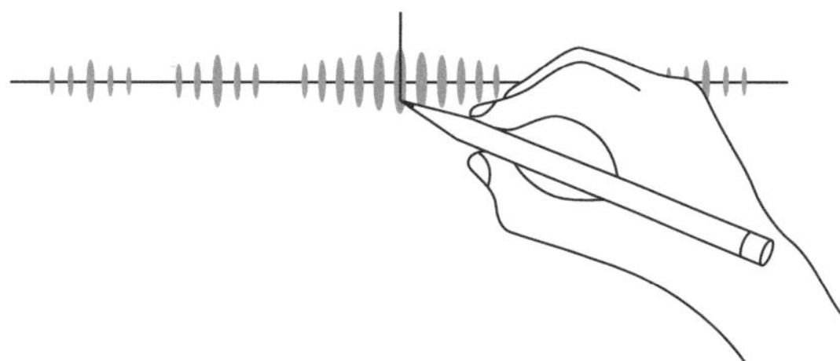

6 Continue to make small marks at the centres of the neighbouring five maxima to the right or left of the central maximum.

7 Turn off the laser, and then record its wavelength, λ, in the Data and analysis section.

The wavelength is usually printed on a sticker on the outside of the laser. If it is not printed on the laser, ask your teacher for this value.

8 Use the measuring tape to measure the distance, L, between the diffraction plate and the white paper. Record this value in the Data and analysis section.

9 Remove the white paper and place it on your lab table.

10 Use the ruler to accurately measure the distance from the zeroth maxima mark ($m = 0$) to each higher order maxima ($m = 1, 2, 3, 4, 5$). Record each distance (x_m) in Table 1 of the Data and analysis section.

DATA AND ANALYSIS

Width between slits: ______________

$\lambda =$ ______________

$L =$ ______________

TABLE 1 Results for maxima spacing

Maximum number, m	Distance from zeroth maximum, x_m (cm)
1	
2	
3	
4	
5	

1 Plot a graph of distance of interference maxima from zeroth maximum (x_m) versus integer order (m) on the blank graph below. Label both axes with the correct scale and units.

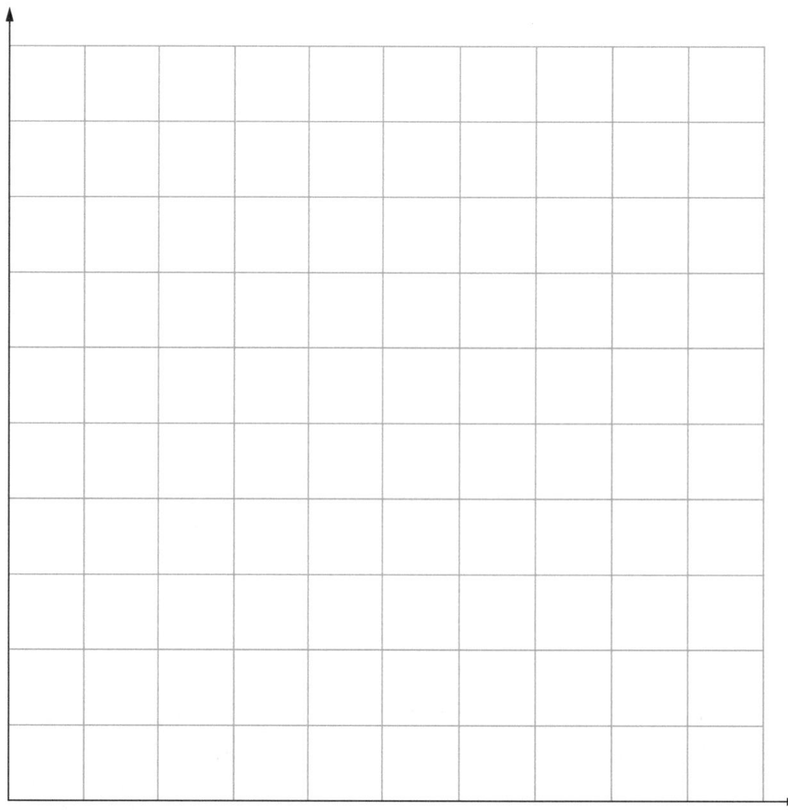

 ISBN 978 1 4886 1936 6

2 Draw a line of best fit through the data in the graph and record the equation of the line in the form $y = mx + c$.

equation of line of best fit: ____________________

3 Use the gradient from the line of best fit to determine the spacing, d, between the parallel slits on the interference plate:

gradient $= \frac{\lambda L}{d}$

slit spacing d (cm): ____________

4 Calculate the percentage difference between your experimental value and the actual value.

$$\text{percent difference} = \left|\frac{\text{actual} - \text{experimental}}{\text{actual}}\right| \times 100\%$$

CONCLUSION

1 What was the experimental value for the spacing between the double slits? How did this compare with the stated value?

2 Based on your calculation of the percentage difference between actual and experimental values, comment on the reliability of conclusions you can draw from this investigation.

3 What factors may have caused error in your experimental value of the slit spacing? Explain how each factor you list could have been avoided or minimised.

4 Explain how your data would differ if you had used slits spaced twice as far apart or half as far apart.

5 Would the data in your experiment differ if the distance between your laser and the double-slit aperture had been much greater? Justify your answer.

RATING MY LEARNING	My understanding improved	Not confident ◄——► Very confident ○ ○ ○ ○ ○	I answered questions without help	Not confident ◄——► Very confident ○ ○ ○ ○ ○	I corrected my errors without help	Not confident ◄——► Very confident ○ ○ ○ ○ ○

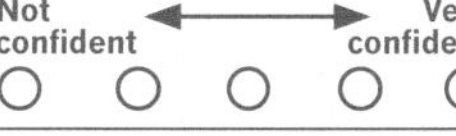

PRACTICAL ACTIVITY 7.5

Polarisation effects

Suggested duration: 55 minutes

INTRODUCTION

Following investigations into two key wave behaviours of light (diffraction and interference), in this activity another of the key behaviours of light suggesting wave behaviour will be explored; polarisation.

A photon of light consists of two wave components: an electric field component, and a magnetic field component. Together they are referred to as a single electromagnetic wave. Polarisation of light occurs when the electrical field component of the electromagnetic wave is constrained to oscillate in one plane. The electric field for polarised light would look like part **(b)** in the figure below; part **(a)** shows unpolarised light. The plane formed by E and the direction of propagation is called the plane of polarisation. Because the wave is travelling along the x-axis (coming out of the page) in the illustration, the plane of polarisation would be the x–y plane in part **(b)**.

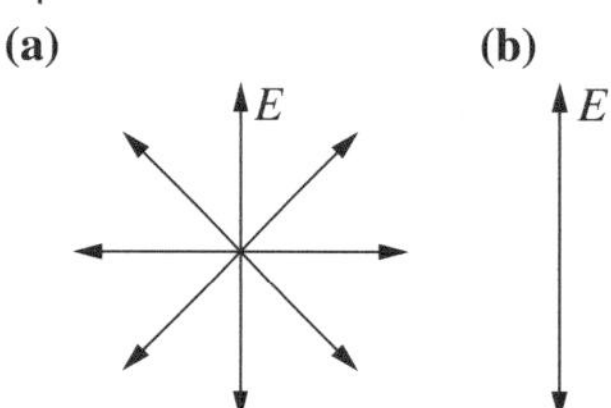

This experiment requires a material that transmits light waves with electric field vectors that oscillate in a particular direction, while absorbing waves with electric field vectors that oscillate outside that direction. Dr Edwin Land developed a suitable material in 1932. He called the material Polaroid. The material is a kind of plastic made of long-chain hydrocarbons. As the sheets cool in the manufacturing process, they are stretched in such a manner as to align the hydrocarbon chains in one direction. Each sheet is dipped into a solution that contains iodine, making the molecules good electrical conductors.

The result is a material that absorbs light waves with electric fields parallel to the hydrocarbon chains, while waves with electric fields perpendicular to the hydrocarbon chains are transmitted.

An unpolarised beam of light incident upon a linear polariser will result in a beam of light that is linearly polarised in the same direction as the polariser's transmission axis. However, the intensity of the resultant beam will be reduced, because all light in the beam not in the same plane as the polariser will be absorbed. If a second polariser (usually called an analyser) is placed in the beam's path after the location of the original polariser, the intensity of the light that is transmitted through it will decrease, again due to the second sheet's absorption. As the angle between transmission axes on both the analyser and polariser gets closer to 90°, the intensity of light transmitted through the analyser will approach zero. This is referred to as Malus' law:

$$I = I_{max} \cos^2 \theta$$

where I is the intensity of the light after passing through the analyser, I_{max} is the intensity of a polarised beam of light incident on an analyser and θ is the angle between the light's polarisation direction and that of the analyser.

MATERIALS

- red laser or laser pointer
- light meter or light sensor and data collection system
- two polarising discs and mounts or polarising sheets and protractors
- optics bench or retort stands and clamps
- black paper and tape

Do not look directly into a laser. Avoid laser beam reflections that may cause eye damage. Be aware of the direction of the laser beam at all times—you may be producing unknown stray reflections.

PURPOSE

To study the effects of polarisation on light intensity, and explore Malus' law.

PROCEDURE

1 Mount the light sensor and connect it to the data-collection system following the manufacturer's instructions. Alternatively, securely clamp a light meter with a retort stand and clamp. Use the black paper to create a tube in front of the light meter to exclude ambient light from around the room.

 ISBN 978 1 4886 1936 6

2 Using either an optics bench or additional retort stands, mount the laser and align it so it is pointed directly at the light meter. Leave sufficient room between the laser and light meter to mount both the polariser sheets. Take and record an initial reading without the polarisers and record this in the Data and analysis section.

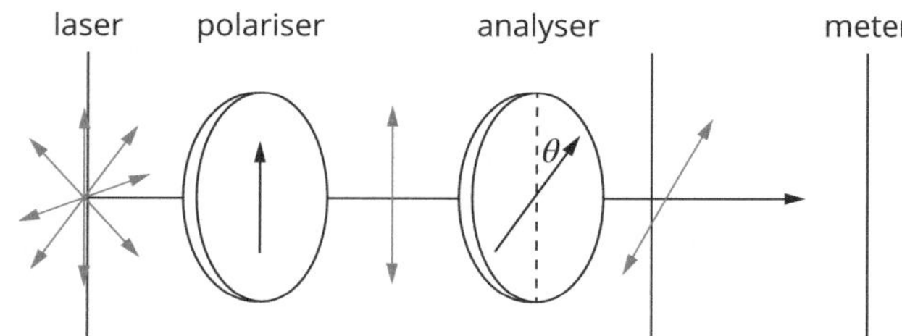

3 Mount and clamp the two polariser discs between the laser and meter or sensor. The first disc will serve as the polariser and the second as the analyser. Rotate the two discs until the transmitted intensity through both discs is greatest. Record this reading in Table 1 in the Data and analysis section. This becomes the zero angle for the angle between polarisers.

4 Rotate the disc closest to the meter by 10°. You can use a protractor to measure the angle. Measure and record the value on the light meter for this second data point.

5 Continue this procedure, changing the angle by 10° each time until you have completed a full revolution (360°) for the analyser. Record the 360° result to confirm that it is the same as the starting reading.

DATA AND ANALYSIS

Initial light level without polarisers: ______________

TABLE 1 Light intensity and angle

Angle (°)	Measured intensity, I	Calculated intensity, $I = I_{max} \cos^2\theta$	% difference
0			
10			
20			
30			
40			
50			
60			
70			
80			
90			
100			
110			
120			
130			
140			
150			
160			
170			
180			
190			
200			

PRACTICAL ACTIVITY 7.5

TABLE 1 *(continued)*

Angle (°)	Measured intensity, I	Calculated intensity, $I = I_{max} \cos^2\theta$	% difference
210			
220			
230			
240			
250			
260			
270			
280			
290			
300			
310			
320			
330			
340			
350			
360			

1 It is important that the distances between polariser, analyser and meter remain exactly the same through out the experiment. Why is the relative distance crucial to the success of the experiment?

2 Why does the point of greatest transmitted intensity through the two polariser discs become the zero angle for the angle between polarisers?

ISBN 978 1 4886 1936 6

3 Draw a graph of measured light intensity versus angle on the axes provided below.

4 Based on the initial value of intensity with the polarisers in place, calculate the expected intensity for each angle.

5 Compare the calculations for intensity to the actual intensities that you measured. To make a quantitative comparison, find the percentage difference and record it in the results table.

$$\% \text{ difference} = \frac{\text{calculated intensity} - \text{measured intensity}}{\text{calculated intensity}} \times 100\%$$

CONCLUSION

1 Comment on your results. Do they support Malus' law with reasonable precision?

2 How could the experimental technique be improved? What was the largest potential source of error?

3 Polarising discs, such as the ones used in this experiment, use long-chain hydrocarbon molecules in the disc material that are made to be electrically conducting. In what direction is the transmission axis for such a disc, relative to the long molecules?

4 Can a sound wave be polarised? Explain your answer.

RATING MY LEARNING	My understanding improved	Not confident ⟷ Very confident ○ ○ ○ ○ ○	I answered questions without help	Not confident ⟷ Very confident ○ ○ ○ ○ ○	I corrected my errors without help	Not confident ⟷ Very confident ○ ○ ○ ○ ○

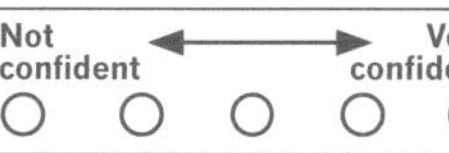

DEPTH STUDY 7.1

Making light work

Suggested time: 3.75 hours

INTRODUCTION

The nature of light under particular conditions can lead to it being seen as more particle-like or more wave-like in behaviour, depending on the circumstances. Historical experiments over many years have refined our understanding of light. Those same historical experiments were also often applied to a practical scenario, or indeed were developed as a solution to a practical problem. This is particularly true where reliable measurements of very small objects are required.

This depth study requires you to investigate how an experiment on the behaviour of light can be used in a practical application. You will develop an inquiry question that requires research or experimentation to develop an informed hypothesis; plan an experimental investigation; analyse primary and secondary data and information from appropriate sources; and use problem-solving techniques to determine the validity of the data and sources. You will evaluate the data to form conclusions by considering the quality of the data. You will process the data and information in order to communicate your findings in a written investigation report of approximately 500–800 words that includes your initial research or planning, experimentation, problem-solving and development of knowledge and understanding of the topic.

Photos of your experiments and modelling should be included to illustrate your progress.

Safety warning:
A number of potential solutions may use lasers. Take care to follow all safety guidelines that apply. Be careful of hot surfaces.

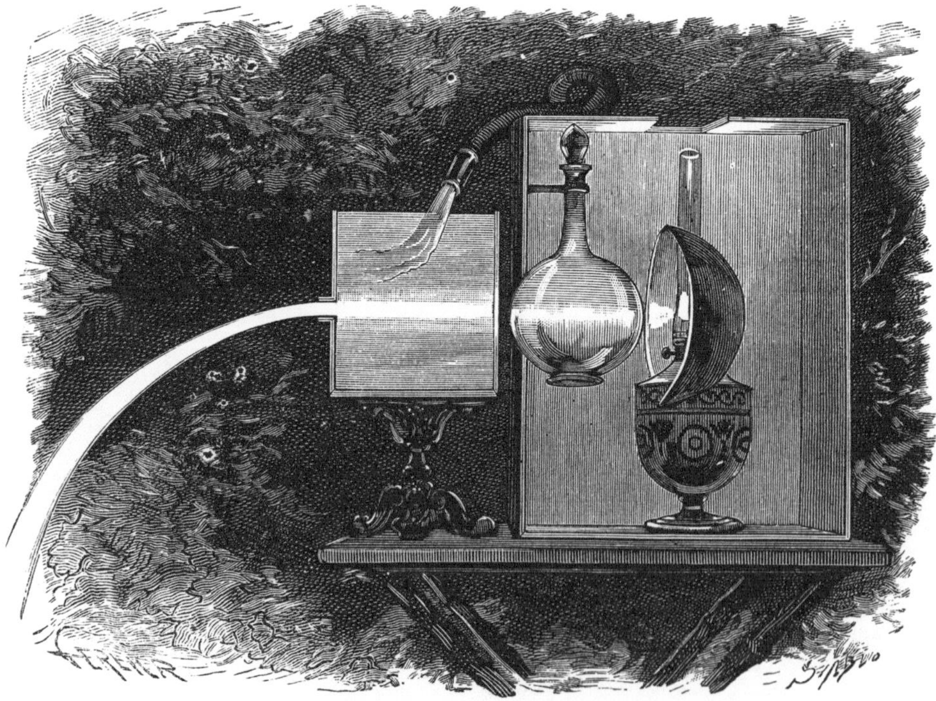

Some historical experiments have more practical applications than others. This 'Light Fountain' from the mid-nineteenth century demonstrated total internal reflection but was seen as largely a novelty at the time. Today, total internal reflection is used in telecommunications networks with the application of fibre-optic cables.

Your report will use appropriate scientific notation and nomenclature, graphs, diagrams and appropriate calculations, and will appropriately apply scientific language that is suitable for the audience and context.

 ISBN 978 1 4886 1936 6

DEPTH STUDY 7.1

PURPOSE

Drawing from the topic suggestions below, research, plan, develop and problem solve an experimental investigation that applies a behaviour of light to solving a practical problem. Your investigation should draw on second-hand data to help you develop a practical method for your experiment, from which you can produce quantifiable first-hand data to analyse.

DESIGN REQUIREMENTS AND CONSTRAINTS

You can work individually or in small groups as resources, time and class constraints allow. Check with your teacher.

Your topic should derive from the application of the behaviour of light to measuring a physical quantity. The following tasks may be expanded upon by your teacher, depending upon time and resource constraints.

- Measure the width of a human hair.
- Find the thickness of a thin film of oil.
- Determine the diameter of a pinhole or small spherical object.
- Find the likely breaking points in a simple implement such as a fork, spoon or spanner.
- Find the points of tension in a glass panel—a windscreen, window or similar.
- Determine the concentration of a sugar solution.
- Determine the temperature of a very hot surface or object.

QUESTIONING AND PREDICTING

1 Why is light suitable for making precise measurements of very small objects?

2 What is the underlying principle or behaviour of light that is applicable to your topic? Conduct some initial research to find historical methods applying a behaviour of light to the task. Describe the principles on which the method is based.

3 Phrase your topic as a question. For example, 'How is diffraction used to find the width of a very thin object?'

4 List three to four possible questions that would help answer your main question. For example, 'What are the limitations of the method?', 'Does it apply to any surface, or are there particular conditions that apply?'

PLANNING YOUR INVESTIGATION

5 Do some initial research around developing a methodology suitable for the resources you have available to you. You may want to rephrase or re-evaluate your questions to make them as clear and specific as possible. Develop a preliminary method and list the resources required. Discuss these with your teacher.

6 State the measurements you will need to take, a reliable method of doing so and the formulae or relationships that will be applied to the analysis.

CONDUCTING YOUR INVESTIGATION

7 Using your initial method from step 5 above, conduct two or more trials. Identify steps where your methodology can be improved.

8 Modify your method and resources. Discuss your prospective changes with fellow students. Describe your final method.

 ISBN 978 1 4886 1936 6

ANALYSING DATA AND INFORMATION

9 Record your data in suitable tables following appropriate scientific conventions and units. You can use a spreadsheet to do so. You may like to print out the sheet and record or list it here as a reference.

Record an estimate of the error in each measurement.

10 Using the relationships you identified, calculate the final quantity (e.g. width of the human hair).

11 Using the estimate of the uncertainty in each measurement, determine an estimate of the error in the final result.

COMMUNICATING

Communicate your findings using a short report. Your report should take the form of an experimental report using the following subheadings:

- Introduction
- Purpose
- Background
- Materials
- Procedure
- Data and analysis
- Conclusion (including your final result with an estimation of the precision of the result)
- Discussion (including answers to the questions below).

12 Look at other students' investigations. Comment on the critical thinking skills, problem solving and scientific processes they employed, which made some more or less effective in determining a reliable final result.

13 What would you change in your final method to make it more reliable? Compare the likely results from your method with the results you expect from these changes.

 ISBN 978 1 4886 1936 6

MODULE 7 • REVIEW QUESTIONS

Multiple choice

1 The emission spectrum of a strange new isotope has been examined. It contains precisely 10 different wavelengths. How many energy levels are in an atom of this isotope?

- **A** 4
- **B** 5
- **C** 10
- **D** 15

2 Two stars are observed apparently close together. The first star has a surface temperature of 5800 K and a colour centred on a wavelength of 500 nm. The second has been found to have a peak colour centred on 580 nm. Its surface temperature is about:

- **A** 5000 K
- **B** 5400 K
- **C** 6700 K
- **D** 5800 K

3 When a proton with a rest mass of 1.67×10^{-27} kg is accelerated from rest, it gains a total energy of 5.0×10^{-10} J. What value of γ does it achieve?

- **A** 3.0
- **B** 3.3
- **C** 0.3
- **D** It cannot be found as special relativity does not apply to accelerated frames of reference.

4 The number of visible nodal lines made by a pair of slits on a screen can be increased by increasing which one or more of the following?

- **A** distance to screen
- **B** wavelength of the light
- **C** separation of the slits
- **D** intensity of the light

5 Which of these types of EMR can be observed to behave most like particles?

- **A** visible light
- **B** radio waves
- **C** ultraviolet rays
- **D** gamma rays

6 Electromagnetic radiation consists of both an electric field and a magnetic field. Which of the following are true?

- **A** The electric and magnetic fields are perpendicular to each other.
- **B** The electric and magnetic fields are parallel but point in opposite directions.
- **C** The electric field oscillates at twice the frequency of the magnetic field.
- **D** Only the electric field oscillates; the magnetic field is constant.

Short answer

7 A student measures the intensity of unpolarised light emitted by an incandescent bulb as $1.80\,\text{W}\,\text{m}^{-2}$ at a distance of 1.20 m. At her disposal she has a number of polarised, untinted filters.

a What intensity would she expect at a distance of 3.6 m from the lamp?

b At the original distance, what intensity would be expected behind one polarised filter?

c A second polarised filter is placed behind the first one at an angle of rotation of 90°. What intensity would be expected now?

d A third sheet is now added behind the first one at an angle of rotation of 45°. What intensity would be expected at the meter now?

8 The graph below shows the factor by which mass increases with increasing velocity (as v approaches the speed of light).

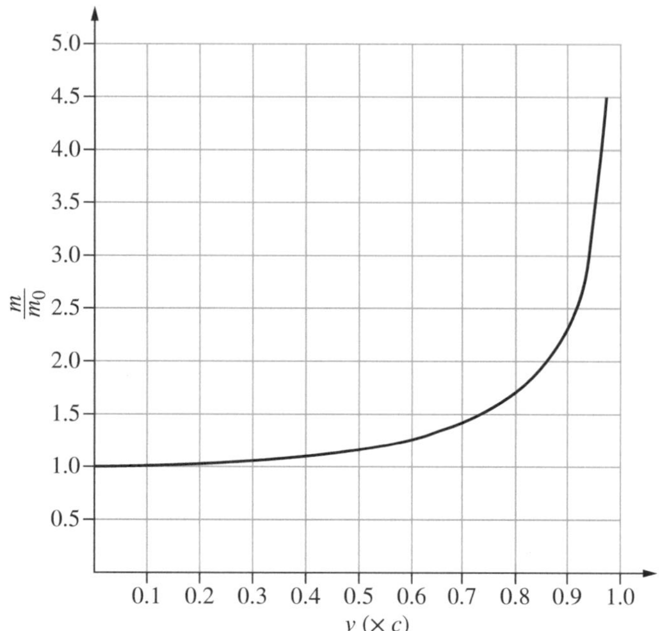

A proton of rest mass 1.67×10^{-27} kg is accelerated in the Large Hadron Collider until it reaches 0.90c.

a Use the graph to estimate the new mass of the proton.

b Use this mass to find how much energy this proton possesses.

c Use the graph to estimate its momentum.

d Explain what is meant by the following statement: 'Trying to accelerate particles just makes them heavier; they don't actually go much faster.'

9 The pattern below was produced by red light of wavelength 630 nm being shone through a pair of slits 0.22 mm apart. The screen on which the pattern was projected was 1.30 m away from the slits.

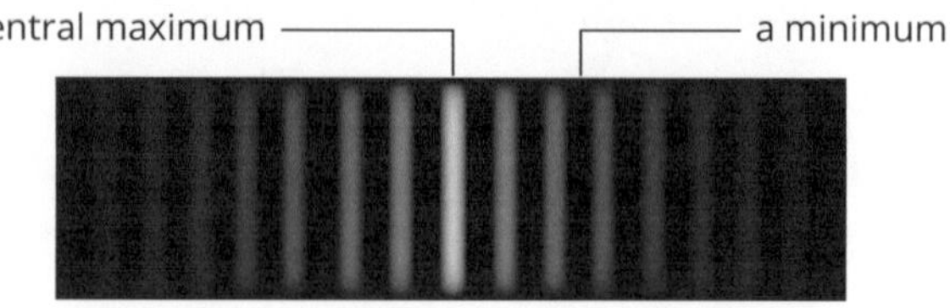

a What is the path difference for the central maximum?

b What is the path difference for the minimum shown?

c What is the distance on the screen from the central maximum to the minimum shown?

d What would happen to the pattern if green light were used instead of red?

10 The energy level diagram for hydrogen is shown below.

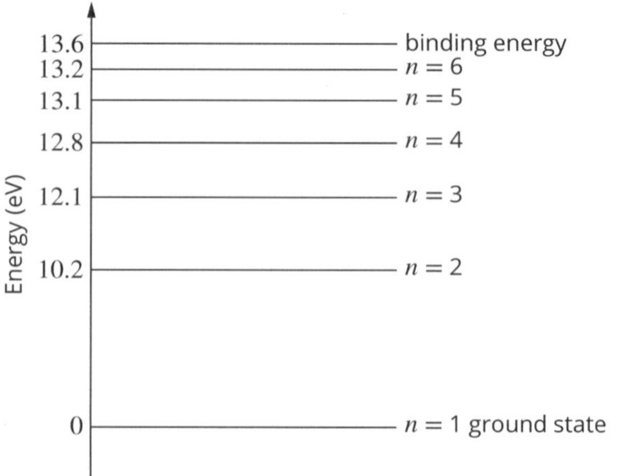

 ISBN 978 1 4886 1936 6

a Calculate the frequency of a photon emitted in a transition from the $n = 5$ level to the $n = 2$ level.

b What wavelength would such a photon have?

c When the absorption spectrum of hydrogen is examined, this particular wavelength is not present. Explain how this occurs.

d What happens when a 12.5 eV photon strikes a hydrogen atom in the ground state?

11 The photoelectric effect is usually cited as strong evidence for a dual nature of light. The diagram shows a typical arrangement used to measure the photoelectric effect.

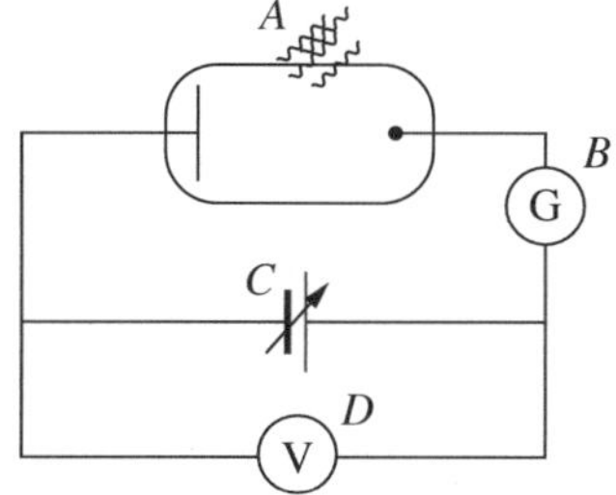

a There are letters identifying the parts of the apparatus. Fill in the blanks in the table below.

Part	Name	What it is for and/or what it does
A	photons/light	
B		
C		
D		

b What factors affect the kinetic energy of the fastest photoelectrons?

c Classical physics predicted that the intensity of the light would increase this energy. Explain how this prediction was made and why it was not observed.

d It is found that the lowest frequency that produces photoelectrons is 5.31×10^{14} Hz. What is the energy of these photons in J?

e What is the size of the work function for the metal in this experiment?

f What will be the energy of the fastest photoelectrons produced by light whose wavelength is 300 nm?

12 Muons with a half-life of 1.56×10^{-6} s are created by cosmic-ray interaction with atoms in the upper atmosphere at a height of 10.0 km. Many of them travel down at 0.98*c* and are detected at the Earth's surface.

a In the Earth's frame of reference, how long would it take a muon to reach the surface?

b How many muon half-lives does this represent?

c Given that after each half-life, only half of the muons remain, estimate how many muons would be expected to arrive at the Earth's surface from an initial population of 10 million?

d It is found that the number of muons detected is much greater than expected. Show how special relativity explained this by time dilation.

e Find the travel time in the muon's frame of reference.

f How many half-lives does this constitute? Out of 10 million muons, how many will reach the surface?

g Outline how this observation can be explained by length contraction, rather than by time dilation.

 ISBN 978 1 4886 1936 6

MODULE 8 From the universe to the atom

Outcomes

By the end of this module you will be able to:

- analyse and evaluate primary and secondary data and information PH12-5
- solve scientific problems using primary and secondary data, critical thinking skills and scientific processes PH12-6
- communicate scientific understanding using suitable language and terminology for a specific audience or purpose PH12-7
- explain and analyse the evidence supporting the relationship between astronomical events and the nucleosynthesis of atoms and relate these to the development of the current model of the atom PH12-15.

Content

ORIGINS OF THE ELEMENTS

INQUIRY QUESTION **What evidence is there for the origins of the elements?**

By the end of this module you will be able to:

- investigate the processes that led to the transformation of radiation into matter that followed the 'Big Bang' CCT ICT
- investigate the evidence that led to the discovery of the expansion of the universe by Hubble (ACSPH138) ICT N
- analyse and apply Einstein's description of the equivalence of energy and mass and relate this to the nuclear reactions that occur in stars (ACSPH031) CCT
- account for the production of emission and absorption spectra and compare these with a continuous black-body spectrum (ACSPH137) CCT ICT
- investigate the key features of stellar spectra and describe how these are used to classify stars N
- investigate the Hertzsprung–Russell diagram and how it can be used to determine the following about a star: CCT ICT N
    - characteristics and evolutionary stage
    - surface temperature
    - colour
    - luminosity
- investigate the types of nucleosynthesis reactions involved in Main Sequence and Post-Main Sequence stars, including but not limited to: CCT ICT
    - proton–proton chain
    - CNO (carbon–nitrogen–oxygen) cycle.

Module 8 • From the universe to the atom

STRUCTURE OF THE ATOM

INQUIRY QUESTION **How is it known that atoms are made up of protons, neutrons and electrons?**

By the end of this module you will be able to:

- investigate, assess and model the experimental evidence supporting the existence and properties of the electron, including: ICT
  - early experiments examining the nature of cathode rays
  - Thomson's charge-to-mass experiment
  - Millikan's oil-drop experiment (ACSPH026)
- investigate, assess and model the experimental evidence supporting the nuclear model of the atom, including: ICT
  - the Geiger–Marsden experiment
  - Rutherford's atomic model
  - Chadwick's discovery of the neutron (ACSPH026).

QUANTUM MECHANICAL NATURE OF THE ATOM

INQUIRY QUESTION **How is it known that classical physics cannot explain the properties of the atom?**

By the end of this module you will be able to:

- assess the limitations of the Rutherford and Bohr atomic models ICT
- investigate the line emission spectra to examine the Balmer series in hydrogen (ACSPH138) ICT
- relate qualitatively and quantitatively the quantised energy levels of the hydrogen atom and the law of conservation of energy to the line emission spectrum of hydrogen using:
  - $E = hf$
  - $E = \frac{hc}{\lambda}$
  - $\frac{1}{\lambda} = R\left[\frac{1}{n_f^2} - \frac{1}{n_i^2}\right]$ (ACSPH136) ICT N
- investigate de Broglie's matter waves, and the experimental evidence that developed the following formula:
  - $\lambda = \frac{h}{mv}$ (ACSPH140) ICT N
- analyse the contribution of Schrödinger to the current model of the atom.

PROPERTIES OF THE NUCLEUS

INQUIRY QUESTION **How can the energy of the atomic nucleus be harnessed?**

By the end of this module you will be able to:

- analyse the spontaneous decay of unstable nuclei, and the properties of the alpha, beta and gamma radiation emitted (ACSPH028, ACSPH030) ICT

- examine the model of half-life in radioactive decay and make quantitative predictions about the activity or amount of a radioactive sample using the following relationships:
  - $N_t = N_0 e^{-\lambda t}$
  - $\lambda = \frac{\ln 2}{t_{1/2}}$

  where N_t = number of particles at time t, N_0 = number of particles present at $t = 0$, λ = decay constant, $t_{1/2}$ = time for half the radioactive amount to decay (ACSPH029) ICT N
- model and explain the process of nuclear fission, including the concepts of controlled and uncontrolled chain reactions, and account for the release of energy in the process (ACSPH033, ACSPH034) ICT
- analyse relationships that represent conservation of mass-energy in spontaneous and artificial nuclear transmutations, including alpha decay, beta decay, nuclear fission and nuclear fusion (ACSPH032) ICT N
- account for the release of energy in the process of nuclear fusion (ACSPH035, ACSPH036) ICT
- predict quantitatively the energy released in nuclear decays or transmutations, including nuclear fission and nuclear fusion, by applying: (ACSPH031, ACSPH035, ACSPH036) ICT N
  - the law of conservation of energy
  - mass defect
  - binding energy
  - Einstein's mass–energy equivalence relationship $E = mc^2$.

DEEP INSIDE THE ATOM

INQUIRY QUESTION **How is it known that human understanding of matter is still incomplete?**

By the end of this module you will be able to:

- analyse the evidence that suggests:
  - that protons and neutrons are not fundamental particles
  - the existence of subatomic particles other than protons, neutrons and electrons
- investigate the Standard Model of matter, including:
  - quarks, and the quark composition hadrons
  - leptons
  - fundamental forces (ACSPH141, ACSPH142) ICT
- investigate the operation and role of particle accelerators in obtaining evidence that tests and/or validates aspects of theories, including the Standard Model of matter (ACSPH120, ACSPH121, ACSPH122, ACSPH146). ICT

Key knowledge

Origins of the elements

THE BIG BANG

Cosmology is the branch of astronomy involving the origin and evolution of the universe. Newton believed that the universe is static and composed of an infinite number of stars that are scattered randomly throughout an infinite space. He believed time and space are steady and are independent of one another. The implication of this statement was: if the universe is static, as Newton described, then why is the sky dark at night? This is known as Olbers' paradox. The sky should be the average brightness of all these stars, as according to Newton the universe existed in a steady state. Einstein overturned part of Newton's theory with his theories of special and general relativity. General relativity predicted an expanding universe, and Edwin Hubble found the evidence that supported these predictions. He saw that the further away a galaxy was, the faster it seemed to be receding from us. This linear relationship is called **Hubble's law**. Hubble's law is expressed by the simple equation

$$v = H_0 d$$

where:

v is the speed of recession (km s^{-1})

H_0 is Hubble's constant ($\text{km s}^{-1}\,\text{Mpc}^{-1}$)

d is the distance away (megaparsecs, where 1 Mpc ≈ 3.3 light years).

Last century, two different theories of the evolution of the universe seemed possible. Either the universe was in a 'steady state' with matter being continuously created, or matter was all created with a 'big bang'. The **steady state theory** suggests that the creation of matter has been occurring continuously, with new matter coming into existence all the time, allowing the density of the universe to remain constant. The steady state theory supports the idea of an 'expanding' universe because matter is being created all the time at just the right rate to keep the density of the universe constant. The **big bang theory** was the main competitor to the steady state theory and is still the leading explanation for how the universe began.

At its simplest, the big bang says the universe as we know it started with a small singularity (a point of infinitely small volume and infinite density), then inflated over the next 13.8 billion years to the cosmos that we know today. This theory follows from the idea that the universe expanded from a very hot and dense state, and as temperatures cooled, matter began to form. Scientists decided that if the big bang did in fact happen, due to this extraordinarily hot and dense state at the beginning, it would have produced large amounts of radiation, as does any hot object. They also proposed that, as the universe expanded, this radiation would have been stretched by the expansion. This left-over radiation was detected and came to be called the cosmic microwave background radiation (CMB). The CMB appears to be mostly smooth, but there are slight variations in temperature that show that matter had started to clump together in the early universe. These clumps of matter formed the galaxies and stars we see today.

The big bang theory predicts many features of the early universe. It is believed that both space and time were created at the big bang, and energy was transformed into matter in an extremely short time called **inflation**. The expansion was so rapid that matter and antimatter pairs of particles could not annihilate until the expansion slowed; usually, matter and antimatter will annihilate when they meet with the release of energy. In the early universe, there was slightly more matter than antimatter, and that excess is the matter that makes up the universe today.

At around 400 million years after the big bang, galaxies and stars began to form. These galaxies and stars formed from matter as the result of very small irregularities in the early universe. The irregularities allowed gravitational attraction to 'pull' matter together. The first atoms to form were hydrogen and helium. The heavy elements necessary for life came later with the explosive supernovae of heavy stars.

THE LIFE CYCLE OF A STAR

Astrophysicists use spectroscopy to analyse light from distant stars, as discussed in Module 7. From spectral analysis, the properties and even the life cycle of stars can be determined. The continuous spectrum (a spectrum containing energy at all wavelengths) of a star provides information about the temperature of its surface—the greater the proportion of blue and UV, the hotter the star. The absorption lines in the star's spectrum tell us what elements are present and under what conditions. The absorption lines in an absorption spectrum correspond to the energies absorbed when electrons are excited to higher levels within the atoms present.

GO TO ➤ Module 7

The spectra of stars are quite similar to the spectrum of the Sun, indicating that they have similar characteristics and processes. Stars are classified according to a system that uses the absorption lines visible in stellar spectra to organise them according to surface temperature (OBAFGKM; hot to cool respectively).

Once data for many stars had been collected it was natural to look for patterns in the data. The **Hertzsprung–Russell (H–R) diagram** plots stars according to surface temperature and colour on the horizontal axis (see Figure 8.1). Surface temperature increases to the left as the colour goes from red to orange then yellow to white and finally blue. Luminosity or absolute magnitude is on the vertical axis and increases upwards. The luminosity scale takes the Sun's value as 1, which corresponds to an absolute magnitude of about +4.8. The absolute magnitude of a star is the magnitude the star would have if it was placed at a distance of 10 parsecs from Earth.

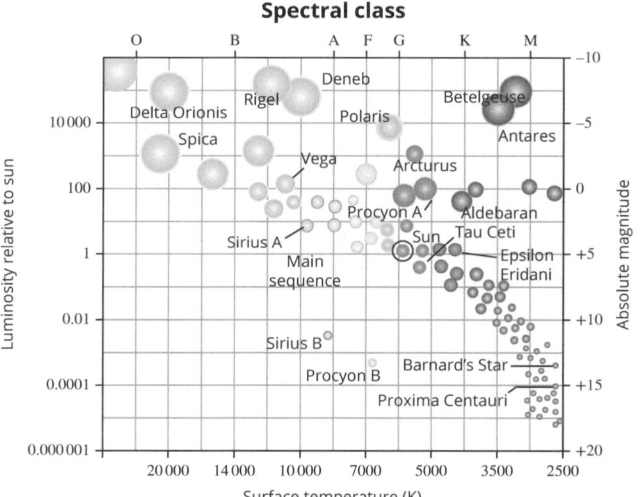

FIGURE 8.1 Hertzsprung–Russell (H–R) diagram

All stars fall into three main groups: **main sequence**; **giants** and super giants; and **white dwarfs**. Moving up the main sequence from right to left on the H–R diagram, the stars become hotter, brighter and larger.

THE LIFE AND DEATH OF STARS

Most stars begin their lives on the main sequence of the H–R diagram, and eventually become giants and then move downwards to become dwarf stars. The Sun, like other main-sequence stars, produces energy by the fusion of hydrogen nuclei to form helium nuclei. Calculations based on the mass and energy output of the Sun show that the amount of energy being produced for each atom in the Sun was around a hundred million times greater than the energy produced by each atom in chemical reactions. This can be explained by mass–energy equivalence and can be found by Einstein's equation:

$$E = mc^2$$

As stars age, the products of nuclear fusion within the star change. Young stars fuse hydrogen to form helium through a proton–proton chain or CNO (carbon–nitrogen–oxygen) cycle. As stars age, they produce increasingly heavier elements. For example, red giants fuse helium into carbon through the triple-alpha process (a set of nuclear fusion reactions by three helium-4 nuclei, i.e alpha particles). Once silicon starts to be fused to produce iron, further nuclear fusion reactions would require a net input of energy and thus do not occur. The star begins to collapse under gravitational forces. Some very large stars end their lives as **supernovae**, which create the heavier elements, allowing the creation of new stars, planets and life itself. Some massive stars collapse to form super-dense **neutron stars**, pulsars and **black holes**.

Structure of the atom

THE ELECTRON

An **electron** is a subatomic particle (particle smaller than the atom) which has a negative electric charge and is denoted by the symbol e. The electron is a fundamental particle because it has no known components or internal structure. The electron was discovered in 1897 by J. J. Thomson when he was studying electric discharges from **cathode ray tubes** (CRTs). A CRT consists of a tube that is vacuum-sealed to keep air out; on either side of the tube are electrodes—the cathode and **anode**. The tube is painted with a phosphor paint, to produce visible light. Thomson noticed that when he applied a high voltage to the electrodes, a beam of light was present throughout the whole length of the tube (shown in Figure 8.2a). He then added a metal plate on either side of the CRT (one positively charged, and the other negatively charged), and he noticed that the ray deflected towards the positive plate. To confirm this discovery, he used a magnetic field, which deflected the beam in a direction that led him to conclude the beam of light must be made of something that is negatively charged (shown in Figure 8.2b).

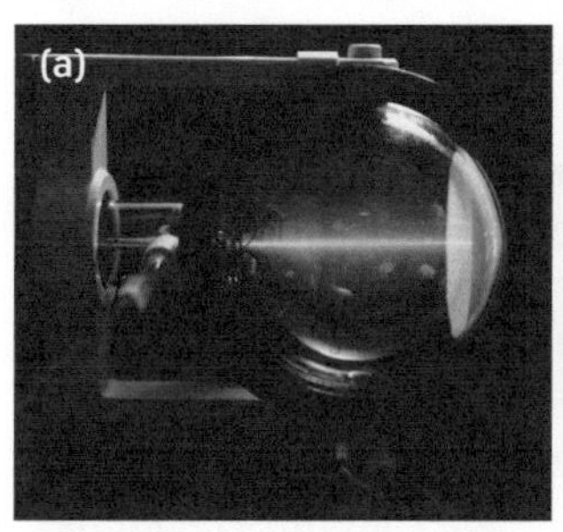

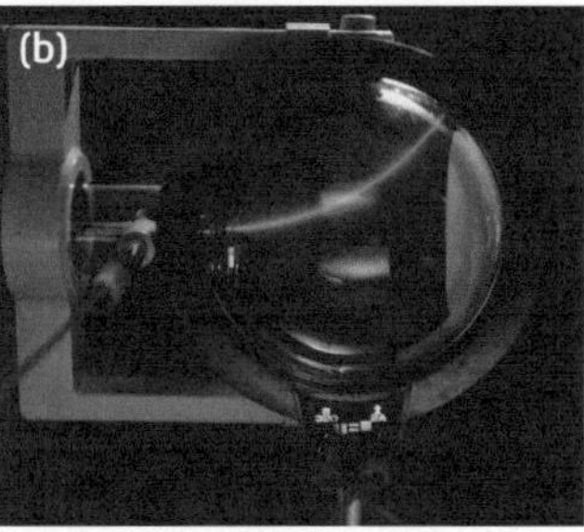

FIGURE 8.2 Cathode ray tube. (a) The straight beam of electrons becomes (b) bent when charged plates are added to the setup and the electrons are deflected.

A cathode ray tube is a useful type of particle accelerator. Electrons are released from a negative terminal, or hot cathode, in a vacuum, and accelerate towards a positive terminal, or anode. As the electrons are accelerated towards the anode they gain velocity and hence kinetic energy. If the electron accelerates from rest ($u = 0$), then this relationship can be described by:

$$K = \frac{1}{2}mv^2 = qV$$

where:

m is the mass (kg)

v is the final velocity (m s^{-1})

q is the charge of the electron (C)

V is the potential difference (V).

This relationship was first discussed in Module 6.

GO TO ➤ page 51

Thomson not only demonstrated that an atom is composed of negatively and positively charged components, but also that the negatively charged component is substantially lighter in mass than the positively charged component. Thomson could not directly measure its mass or charge, but the degree of deflection of the beam could reveal the ratio of the charge of the particle to its mass, e/m. (In this case, the charge is represented by e, as the charge of the electron, rather than by the usual symbol for charge, q.)

Thomson applied a pair of magnets and an electrical potential at right angles to each other through which the cathode ray passed, and this caused forces on the particles in the beam in opposite directions. Recall from Module 6 that a current moving perpendicular to a magnetic field will create a force. If a moving charge experiences a force of constant magnitude that remains at right angles to its motion, its direction will be changed but not its speed. In this way, bending magnets act to alter the path of the electron beam, rather than to speed the electrons up. Charged particles in a magnetic field move in a circular path described by:

$$r = \frac{mv}{qB}$$

where:

r is the radius of the path (m)

m is the mass of the particle (kg; for an electron $m_e = 9.109 \times 10^{-31}$ kg)

v is the speed of the charge (m s^{-1})

q is the charge on the particle (C; for an electron $e = 1.602 \times 10^{-19}$ C)

B is the strength of the magnetic field (T).

By altering the fields such that the beam was undeflected, the velocity of the particles would be the ratio of the electrical to the magnetic field. Then by removing the magnetic field, the particles' motion is affected only by the electrical field, which can then be used to find the ratio.

Thomson's work superseded the Dalton atomic model (Dalton thought atoms to be literally 'atomos' meaning 'indivisible') with the plum pudding atomic model. Thomson suggested that atoms were a cloud of positive charge containing evenly spread negative electrons. Two physicists, Robert Millikan and Harvey Fletcher, later created the oil-drop experiment to measure the charge on the electron. By suspending a charged oil drop using an electrical potential to balance the downwards force due to gravity ($F_E = F_g$), the charge can be found using

$$qE = mg$$

where:

q is the charge on the oil drop (C)

E is the strength of the electric field (N C^{-1})

m is the mass of the oil drop (kg)

g is the acceleration due to gravity (N kg^{-1}).

Millikan's oil-drop experiment quantified the charge on the electron as -1.6×10^{-19} C, and therefore provided the mass of the electron: 9.1×10^{-31} kg.

These results also showed that atoms were composed of smaller components, and overturned the Dalton atomic model that atoms are indivisible.

NUCLEAR MODEL OF THE ATOM

In 1899, a physicist by the name Ernest Rutherford demonstrated that there were at least two distinct types of radiation: alpha radiation (a helium nucleus with no electrons) and beta radiation (an electron or positron). Rutherford studied under J. J. Thomson, and found that radiation could be classified according to its penetrating ability—the alpha (α) particle, which is easily absorbed, and the beta (β) particle, which is more penetrating. Rutherford had found earlier that alpha particles can be deflected in electrical and magnetic fields and their positions measured on a photographic film. However, when the alpha particles passed through a thin sheet of the mineral mica, the film images were blurred. Rutherford's associate Hans Geiger improved the experiment by replacing the mica sheet with gold foil, which should give a larger deflection (Figure 8.3).

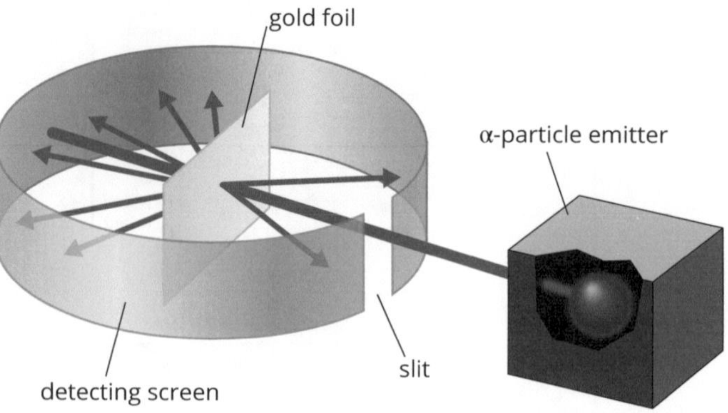

FIGURE 8.3 The gold-foil experiment

In 1909, Geiger and his student Ernest Marsden explored the idea of looking for alpha particles scattered at large angles. Most of the alpha particles went straight through the gold foil and some scattered at different angles, but a few bounced straight back (shown in Figure 8.4b). If Thomson's model was accurate, the alpha particles would travel through the atom, affected only weakly by electrical forces (Figure 8.4a). The Geiger–Marsden scattering experiment showed the mass of the atom is not distributed throughout the space occupied by the atom but rather constrained to a nucleus. Rutherford proposed a nuclear model of the atom, to supersede Thomson's plum pudding model. The nuclear model consisted of positive charges concentrated in the massive centre of the atom, which he called the nucleus, with electrons in orbits around it (shown in Figure 8.4c). However, Rutherford's nuclear model did not explain the electron arrangement—an electrically charged particle circling a centre would continually lose energy as electromagnetic radiation, but this is not the case, as atoms are stable. It also did not explain how positive charges in the nucleus did not mutually repel.

Rutherford also recognised another problem with the nuclear model. If the nucleus contains positively charged protons, electrical repulsion should not allow stable atomic nuclei. In 1931, Walther Bothe and Herbert Becker in Berlin observed that when alpha particles impinge on the light elements lithium, beryllium or boron, a new kind of radiation emerged. This radiation was unusually penetrating and was unaffected by an electric field. The assumption was that the emissions were gamma rays (the most penetrating radiation—discussed later on in this module). James Chadwick, one of Rutherford's former students, did not agree that Bothe's radiation was gamma rays, as he had studied gamma rays as a student and knew that protons are too heavy to be dislodged by gamma rays. Chadwick correctly interpreted the Bothe–Becker and Joliot–Curie experimental results as a demonstration of the neutron. The **neutron** has no charge, so could not be investigated through the interactions of charged particles with electric fields that had led to the discovery of the proton and electron. Chadwick calculated that the particle had a mass close to the proton's mass, but without an electric charge. Subsequent investigations confirmed his experiments.

With Chadwick's announcement, models for atomic nuclei composed of protons and neutrons soon followed.

 ISBN 978 1 4886 1936 6

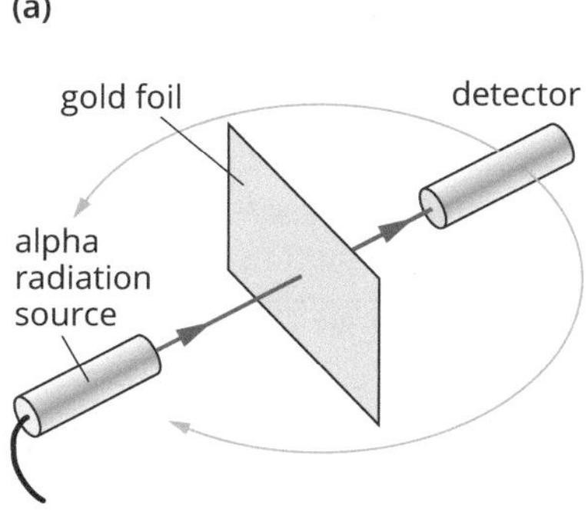

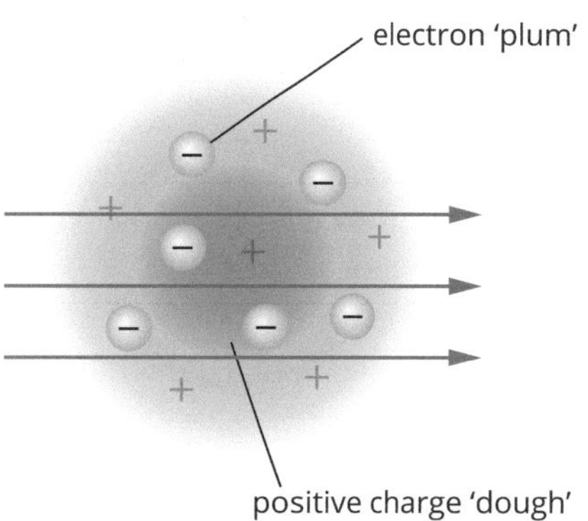

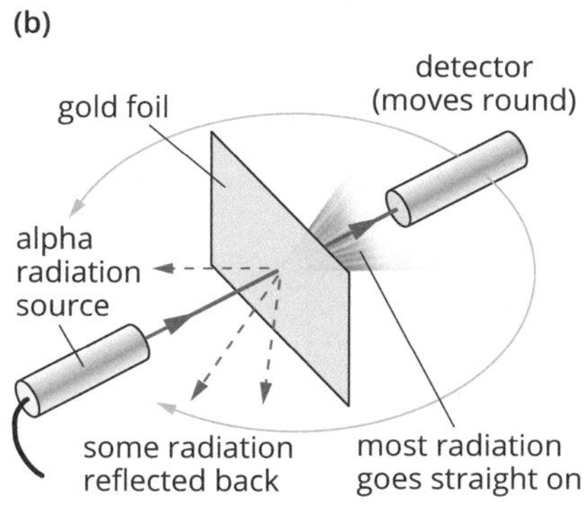

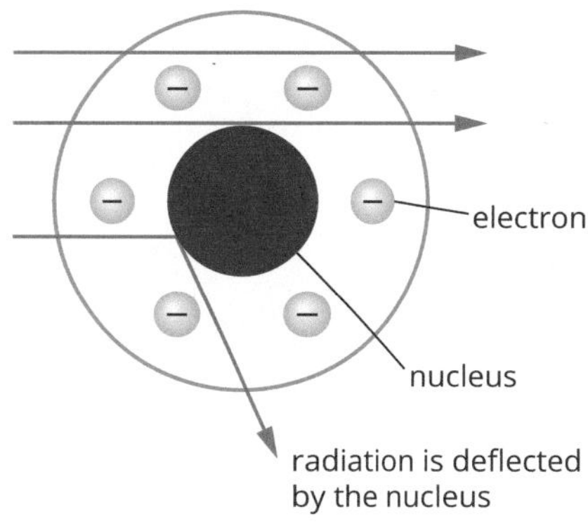

FIGURE 8.4 (a) Thomson model, and (b) Rutherford model

Quantum mechanical nature of the atom

BOHR MODEL

The work of Thomson and Rutherford established the nuclear model of the atom—a positive nucleus surrounded by electrons. But these models had their limitations. For example, Rutherford's model proposed that electrons revolve around the nucleus in circular orbits. According to the law of conservation of energy, the orbiting electrons should emit energy due to their motion. That energy loss would in turn cause their circular orbits to shrink.

A Danish physicist by the name of Niels Bohr published a solution to this instability problem.

1. Electrons can exist only in certain energy states, and that while they are in this state the electrons do not radiate energy.
2. When an electron undergoes a transition from an upper energy state to a lower energy state, a photon of energy is emitted.

Bohr proposed that the quantum mechanical nature of matter was a better way to understand the structure of the atom.

Bohr applied Planck's quantum theory to Rutherford's model, and suggested that electrons in atoms orbit the nucleus in specially defined energy levels, and no radiation is emitted or absorbed unless the electron can jump from its energy level to another. Energy levels can be shown as a number of horizontal lines on a graph. The graph in Figure 8.5 shows the energy levels for sodium gas where $n = 1$ is the ground state (the lowest energy state of an atom) and the highest level is where ionisation occurs and the electron escapes the atom.

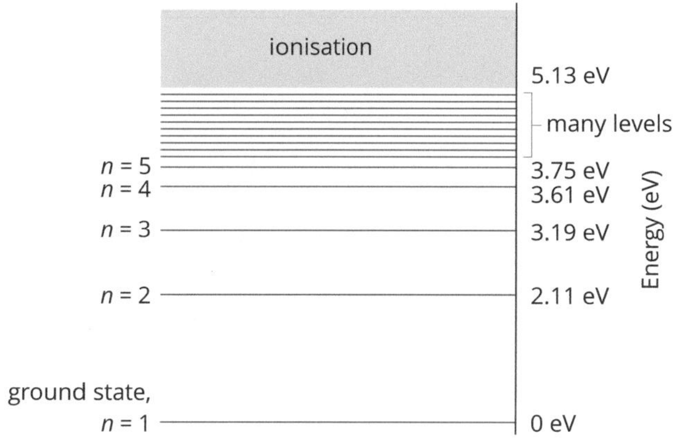

FIGURE 8.5 Energy levels for sodium gas

Bohr's main ideas were as follows.

- The electron moves in a circular orbit around the nucleus of an atom.
- The force keeping the electron moving in a circle is the electrostatic force of attraction (the positive nucleus attracts the negative electron).
- A number of allowable orbits of different radii exist for each atom and are labelled $n = 1, 2, 3$, etc. The electron may occupy only these orbits.

- An electron ordinarily occupies the lowest-energy orbit available.
- An electron can jump to a higher energy level by absorbing some energy. The absorbed energy must be exactly equal to the difference between the electron's initial and final energy levels.
- Electromagnetic radiation is emitted by an excited atom when an electron falls from a higher energy level to a lower energy level. The energy of the emitted light will be exactly equal to the energy difference between the electron's initial and final levels.

The last two points of Bohr's idea suggest that electron energies are quantised, since only certain values are allowed, further supporting Planck's model. As discussed in Module 7, an emission spectrum shows only certain colours or specific wavelengths of light for a certain element. Bohr proposed that an emission spectrum is produced by energised atoms as electrons undergo transitions between energy levels. An electron that drops between energy levels emits a photon (light) of energy equal to the difference between the energy levels. The energy of the photon then determines the colour of the light. The production of spectra suggests an internal structure to the atom. Recall from Module 7 that a line emission spectrum is produced by energised atoms, and an absorption spectrum is created when white light passes through a cold gas. The spectrum for any particular element is unique to that element.

GO TO ➤ page 101

Bohr labelled the possible electron orbits for the hydrogen atom with a quantum number (n), and he was able to calculate the energy associated with each quantum number. Using these energy levels, he could theoretically predict the wavelengths of all of the lines of the hydrogen emission spectrum using Planck's equation:

$$E = \frac{hc}{\lambda}$$

where:

h is Planck's constant (6.626×10^{-34} J s)

c is the speed of light (3.0×10^{8} m s^{-1})

λ is the wavelength (m)

The frequency, f, of a photon emitted or absorbed by a hydrogen atom can be calculated from the difference between the energy levels involved:

$$\text{i.e. } E_2 - E_1 = hf \text{ or } \frac{hc}{\lambda}$$

In 1885, the Swiss mathematician Johann Balmer found an empirical equation that predicted the wavelength of the visible lines of the hydrogen emission spectrum. Later, in 1888, Johannes Rydberg realised that Balmer's formula was a special case of the more general formula:

$$\frac{1}{\lambda} = R\left[\frac{1}{n_f^2} - \frac{1}{n_i^2}\right]$$

where R is the Rydberg constant for hydrogen (1.097×10^{7} m^{-1}) and n_f and n_i are any two integers where $n_i > n_f$. This equation predicted that there should be spectral lines in other parts of the electromagnetic spectrum. The ultraviolet series was later observed by Theodore Lyman, and two different infrared series were observed by Friedrich Paschen and Frederick Brackett. The energy transitions for these series—Balmer, Lyman and Paschen—are shown in Figure 8.6.

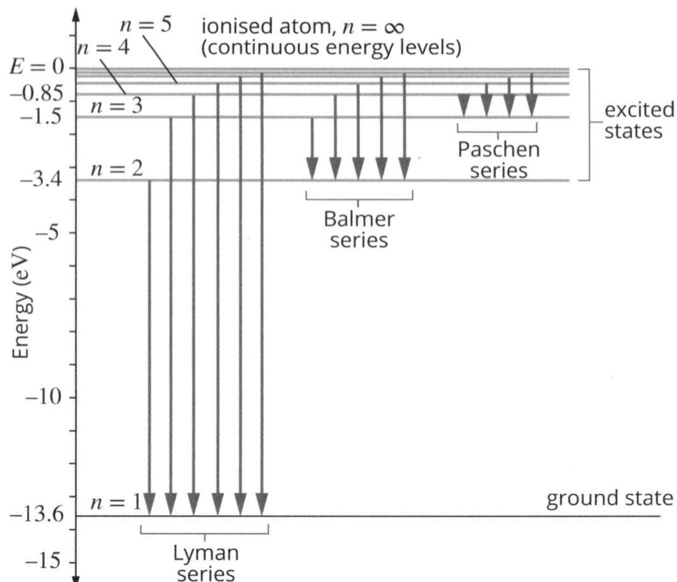

FIGURE 8.6 The energy transitions for the Balmer, Lyman and Paschen series.

The Bohr model of the atom is limited in its application, but was a significant development at the time as it took a quantum approach to the energy levels of atoms and incorporated the quantum nature of electromagnetic radiation. It could only really be accurately applied to single-electron atoms—hydrogen and ionised helium. It modelled inner-shell electrons well but could not predict the higher-energy orbits of multi-electron atoms. Nor could it explain the discovery of the continuous spectrum emitted by solids. A more complex quantum approach was required.

QUANTUM MODEL OF THE ATOM

The wave and particle models for light seem fundamentally incompatible. Waves are continuous and are described in terms of wavelength and frequency. Particles are discrete and are described by physical dimensions such as their mass and radius. In 1924, French physicist Louis de Broglie proposed a ground-breaking theory and suggested that since light demonstrates both wave- and particle-like properties, then perhaps matter might sometimes demonstrate wave-like properties. He quantified this theory by predicting that the wavelength of a particle would be given by the de Broglie equation:

$$\lambda = \frac{h}{p} = \frac{h}{mv}$$

where:

λ is the wavelength of the particle (m)

h is Planck's constant (6.626×10^{-34} J s)

p is the momentum of the particle (kg m s^{-1})

m is the mass of the particle (kg)

v is the velocity of the particle (m s^{-1}).

 ISBN 978 1 4886 1936 6

It should be noted that when this equation is used for everyday objects such as a cricket ball (with a mass of around 60 g), the wavelength is extremely small (about 9.9×10^{-35} m) and is not noticeable. However, on the atomic level, energy and matter exhibit the characteristics of both waves and particles.

All matter, like light, has a dual nature. Through everyday experience matter is particle-like, but under some situations it has a wave-like nature. This symmetry in nature—the dual nature of light and matter—is referred to as **wave–particle duality**. For example, Young's double-slit experiment is explained by a wave model but produces the same interference and diffraction patterns when one photon at a time or one electron at a time is passed through the slits. Figure 8.7 illustrates that an interference pattern can be built up over time by a series of single photons passing through an apparatus like that used in Young's experiment, demonstrating the wave–particle duality of light.

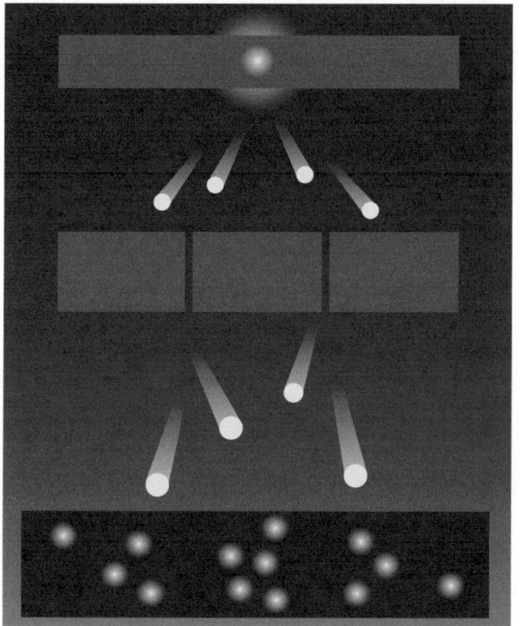

FIGURE 8.7 Young's double-slit experiment using a series of single photons

De Broglie viewed electrons as matter waves. His standing-wave model for electron orbits provided a physical explanation for electrons only being able to occupy particular energy levels in atoms. He suggested that the only way that the electron could maintain a steady energy level was if it established a **standing wave**. The electrons moving in this circular standing-wave pattern can only be established if the circumference of the circle is equal to a whole number of wavelengths.

 A standing wave forms because of the perfectly timed interference of two waves passing through the same medium. The observed wave pattern appears to have particular points that are standing still.

In 1925, the Austrian physicist Erwin Schrödinger built on the work of Niels Bohr by developing a mathematical equation that could describe the wave behaviour of electrons in situations other than the simple hydrogen atom. In Schrödinger's model, the wave properties of electrons are interpreted as representing the probability of finding an electron in a certain location. Quantum mechanics is the name given to the study of the wave properties of electrons.

Heisenberg's uncertainty principle results from wave–particle duality and states that it is not possible to know the exact position and momentum of a particle simultaneously. The more exactly the position of a subatomic particle is known, the less is known about its momentum. Similarly, the more precisely the momentum of a particle is measured, the less certain is its exact position. The Heisenberg uncertainty principle describes a limit to which some quantities can be measured. The uncertainty principle applies a limit to the simplified idea of electrons as particles—i.e. that the position and the velocity of an electron cannot both be known at exactly the same time, and that the amount of energy at a particular time, t, is also uncertain.

The single-slit diffraction experiment provides an example of the application of the uncertainty principle. The diffraction pattern is not produced by photon interactions but rather the probability of where a single photon may end up. Some positions are more likely than others, hence there is a larger central maximum intensity (see Figure 8.8).

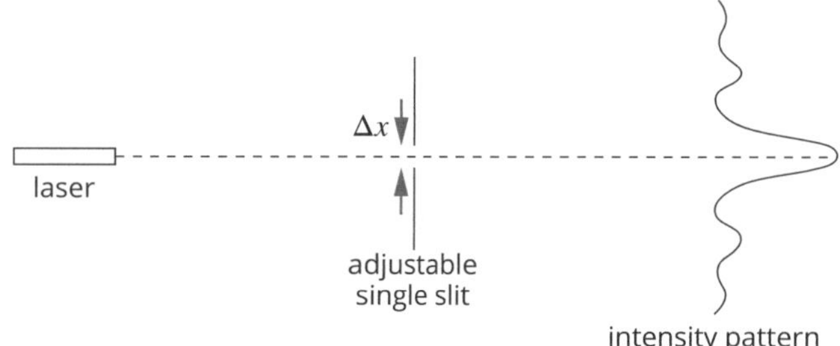

FIGURE 8.8 Single-slit diffraction experiment and the intensity pattern produced

In the quantum mechanical view of the atom, electrons are not particles and do not have clearly defined orbits. Their paths, due to their wave nature, can be explained as a probability distribution of their positions as particles or an electron cloud.

Properties of the nucleus

RADIOACTIVE DECAY

The nucleus of an atom consists of positively charged protons and neutral neutrons. Collectively, protons and neutrons are known as nucleons. Negatively charged electrons then surround the nucleus. In this structure, as protons are almost touching, the electrostatic repulsive force is enormous. Neutrons add distance between these protons and also act as a 'glue'. This 'glue' is the 'strong force' which binds the nucleons together. Nuclear stability is the result of the strong nuclear force between nucleons over a very short distance, which opposes electrostatic repulsion between protons. Atoms with unstable nuclei break down to form more stable nuclei. This is radioactive decay.

The nucleus of the atom is extremely small but contains most of the atom's mass. A particular atom can be represented and identified by using an atomic symbol. This includes the atomic number, Z, which is the number of protons in the nucleus, and the mass number, A, which is the number of nucleons in the nucleus, i.e. the combined number of protons and neutrons. Elements are represented as ${}^{A}_{Z}X$; for example, hydrogen is represented as ${}^{1}_{1}H$. **Isotopes** of an element have the same number of protons but different numbers of neutrons.

Isotopes of an element are chemically identical to each other, but have different physical properties. For example, water made with hydrogen atoms with two neutrons is 'heavier', and hence has a higher density and boiling point than water made with hydrogen with one neutron. An unstable isotope—a radioisotope—may spontaneously decay by emitting alpha, beta or gamma radiation from its nucleus. Alpha, beta and gamma radiation are the three types of nuclear radiation and are classified according to their penetrating ability.

An **alpha particle** (α) consists of two protons and two neutrons and is emitted from the nuclei of some radioisotopes. It is identical to a helium nucleus and can be written as ${}^{4}_{2}He$. Alpha (α) particles are ejected from the nucleus at around 10% of the speed of light. They have a double positive charge (more chance of ionising) and are relatively heavy, which consequently gives them poor penetrating power. This radiation can barely penetrate paper.

Beta (β) particles emanate from the nucleus at up to 90% of the speed of light. They are much lighter than alpha particles. A beta-minus particle (β^- or ${}^{0}_{-1}\beta$) is an electron that has been emitted from the nucleus of a radioactive atom as a result of a neutron transmutating (when one particle changes into another) into a proton. An antineutrino is always emitted with a beta-minus particle. A beta-plus particle (β^+) or positron is a positively charged electron (${}^{0}_{+1}\beta$) that has been emitted from the nucleus of a radioactive atom as a result of a proton transmutating into a neutron. A neutrino is always emitted with a beta-plus particle. Beta-minus particles have a single negative charge while beta-plus particles have a single positive charge. Beta radiation has a moderate penetrating ability, due to its negative charge.

The most penetrating radiation is a **gamma ray**, γ. This is high-energy electromagnetic radiation that is emitted from the nuclei of radioactive atoms. Gamma rays always accompany an alpha or a beta emission. As they are electromagnetic radiation, they travel at the speed of light and have no charge. Gamma radiation can penetrate several centimetres of lead, in comparison to beta, which can penetrate a few millimetres of aluminum.

HALF-LIFE

It is impossible to predict which individual nuclei will decay in a given time, but it is possible to predict the number of nuclei that will decay in a given time from a particular source. The **half-life** of a radioisotope describes how long it takes for half of the nuclei in a given mass to decay.

 The half-life ($t_{1/2}$) of a radioisotope is the time that it takes for half of the nuclei of the sample radioisotope to decay.

The count rate is the **activity** of the sample. The activity of a sample indicates the number of emissions per second. Activity is measured in becquerels (Bq), where 1 Bq = 1 emission per second.

The number of atoms of a radioisotope will decrease over time. Over one half-life, the number of atoms of a radioisotope will halve. The half-life equation can be used to calculate the activity (A) or number (N) of a radioisotope remaining after a number of half-lives (n) has passed:

$$N = N_0\left(\frac{1}{2}\right)^n \text{ and } A = A_0\left(\frac{1}{2}\right)^n$$

where N_0 and A_0 represent the initial number and activity of radioactive nuclei respectively.

Alternatively, the number of particles remaining at a point in time can also be found by using the following formulae:

$$N_t = N_0 e^{-\lambda t}$$

$$\lambda = \frac{\ln 2}{t_{1/2}}$$

where:

N_t is the number of particles at time t

N_0 is the initial number of particles

λ is the decay constant

$t_{1/2}$ is the time for half the radioactive amount to decay (s).

The decay constant is the fraction of the number of atoms that decay in 1 s.

When a radionuclide decays, its daughter nucleus is usually itself radioactive. This daughter will then decay to a grand-daughter nucleus, which may also be radioactive, and so on. This is called a **decay series**.

Carbon dating is a technique used by archaeologists to determine the ages of fossils and ancient objects that were made from plant matter. This method involves measuring and comparing the proportion of two isotopes of carbon, ${}^{12}C$ and ${}^{14}C$, in the specimen. All living plants take in carbon dioxide, a small amount of which contains the radioactive isotope ${}^{14}C$. When plants or animals die, this ${}^{14}C$ decays, so the ratio of ${}^{14}C : {}^{12}C$ drops. The ${}^{14}C : {}^{12}C$ ratio is used to determine the time that has elapsed since the death of a plant or animal.

NUCLEAR FUSION AND FISSION

Within a nucleus, forces of attraction and repulsion are acting. The long-distance electrostatic force of repulsion acts between the protons. The short-distance strong nuclear force of attraction acts between every nucleon. **Nuclear fission** and **nuclear fusion** are the two different types of reactions that release energy.

Nuclear fission occurs when a nucleus is made to split and release a number of neutrons. It is also sometimes defined as the process of splitting an atomic nucleus into fission fragments. This can be induced by striking a fissile nucleus with a neutron. A relatively large amount of energy is released during this process. Fissile nuclei are generally heavy atoms with large numbers of nucleons. The nuclei of such heavy atoms are struck by neutrons, initiating the fission process, or they can also occur due to electrostatic repulsion created by the large numbers of protons within the nuclei of heavy atoms.

Nuclear fusion is the combining of light nuclei to form heavier nuclei. Extremely high temperatures are required for fusion to occur, to overcome the electrostatic repulsion between the protons in the nuclei. A small amount of mass is lost during a fusion reaction, and converted into **binding energy**; this was one of Einstein's greatest discoveries.

 Binding energy is the energy that binds or holds the nuclear particles together. This energy overcomes the electrostatic repulsive forces.

Mass and energy are two forms of the same thing; neither can be created nor destroyed but mass can be converted into energy (using Einstein's mass–energy equivalence, $E = mc^2$). Nuclei are positively charged and so repel each other due to electrostatic forces. Approaching nuclei must have enough speed to overcome the electrostatic forces and get close enough for the strong nuclear force to take effect. The energy that is required to achieve this is called the energy barrier. Fusion reactions do not occur naturally on our planet but are the principal type of reaction found in stars. The large masses, densities and high temperatures of stars provide the initial energies needed to fuel fusion reactions. The amount of energy released per nucleon is greater for fusion than for fission processes.

ENERGY FROM NUCLEAR REACTIONS

When fission occurs, the mass of the fission fragments is always less than the mass of the original particles; this is known as the **mass defect**. This decrease in mass is proportional to its binding energy. Binding energy indicates how much energy is needed to separate the nucleus into its individual protons and neutrons. Einstein showed that mass and energy are equivalent and can be converted to one another. If a large nucleus such as uranium splits into two fragments, the binding energy per nucleon increases. This is the energy released during fission.

The energy released, E, is given by Einstein's equation:

$$E = mc^2$$

where:

c is the speed of light ($3.0 \times 10^8 \text{ m s}^{-1}$)

m is the mass (kg).

The energy of subatomic particles is usually measured in electron-volts (eV), where $1 \text{ eV} = 1.602 \times 10^{-19} \text{ J}$.

 To convert from eV to J: multiply by 1.602×10^{-19}.

To convert from J to eV: divide by 1.602×10^{-19}.

As already mentioned, a small amount of mass is also lost during a fusion reaction. This mass is also related to the energy produced according to $E = mc^2$. The binding energy per nucleon increases dramatically for the very small nuclei. As nucleons fuse together, the binding energy per nucleon increases. This is the energy released during fusion.

Deep inside the atom

THE STANDARD MODEL

Our understanding of the universe progresses by both theory and experimentation. New particles are sometimes predicted before any experimental evidence is available, while sometimes observations in experimental events lead to the discovery of a new particle. The **Standard Model** is the name given to the theory of fundamental particles and how they interact. This model of particle physics explains three of the four fundamental forces in the universe (electromagnetism, the strong nuclear force and the weak nuclear force) in terms of an exchange of particles called **gauge bosons**. Gauge bosons are one of the two fundamental classes of particles, the other being fermions.

Previously, forces were thought of as being exerted on particles by fields; however, in the Standard Model this is resolved and forces are thought to be exerted through the exchange of other particles. An example of this is to consider two inline skaters who are initially stationary, standing facing each other, and then begin to pass a football back and forth to each other. As they do this, they will begin to move away from each other. This is due to the conservation of momentum each time they throw and catch the ball. This situation could be likened to two particles experiencing a repulsive (pushing away) force. Each fundamental force has its own boson. The gauge bosons for these three forces are the photon for the electromagnetic force, the gluon for the strong nuclear force, and the Z, W^- and W^+ for the weak nuclear force.

The other fundamental class of particle is called the fermion, which makes up all matter. **Fermions** are divided into two groups of particles called **quarks** and **leptons**, with a total of 12 fundamental particles. Six of these 12 particles are quarks and combine in groups of two or three to form the hundreds of known particles. The six different types of quarks are referred to as the six 'flavours' of quarks, and are up (u), down (d), strange (s), charmed (c), bottom (b) and top (t). Quarks experience the strong nuclear force mediated by gluons (this is what separates them from the leptons) and combine to form **hadrons**. Quarks have only been found bound into hadrons; they cannot exist alone. Hadrons can be further categorised into **baryons**, which are made of three quarks (and include the proton and neutron), and mesons, which are made of two quarks (a quark–antiquark pair). Figure 8.9 illustrates the proton and neutron, and shows that the proton is made of two up quarks and one down quark (uud) and the neutron is made of one up quark and two down quarks (udd). Each of the three quarks in a baryon must have a different colour charge of red, green or blue and must always bond together to give a total colour charge of 'white' (red + blue + green).

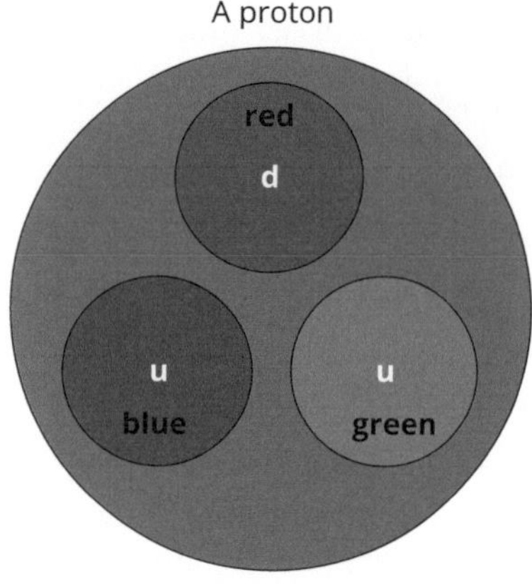

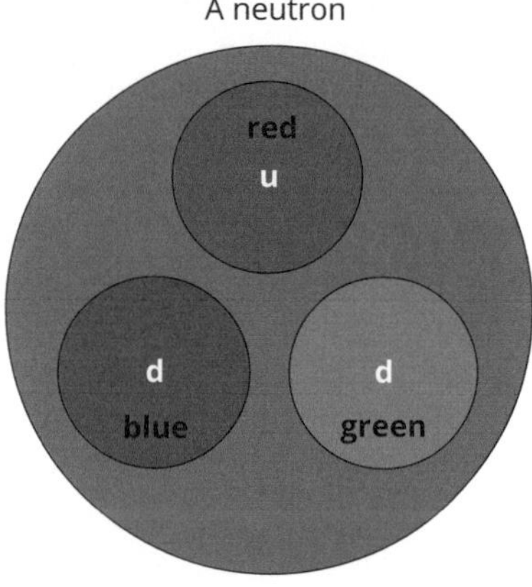

FIGURE 8.9 Baryons: the proton and the neutron

Leptons are a group of particles that interact by exchanging W and Z bosons, which mediate the weak nuclear force. Electrons are in this group of particles. These particles also include the muon and tau particles, as well as their corresponding neutrinos and the antimatter opposites of these six particles.

Antimatter particles have similar properties, like mass, spin and lifetime, to their corresponding particles. Their electric charge, and other characteristics called quantum numbers, are the same in magnitude but have an opposite sign. When a matter particle and its antimatter particle collide, they completely annihilate to produce a photon with the equivalent energy of the two particles, calculated by Einstein's mass–energy equation $E = mc^2$.

EVIDENCE FOR THE STANDARD MODEL

Until the 1960s, physicists did not know how particles got their mass. Using the Standard Model, they predicted that there was a yet-undiscovered particle that would interact with other particles and give them what is measured as mass. This particle came to be known as the **Higgs boson**. The search for this, and for other particles, led physicists and engineers to collaborate to build the **Large Hadron Collider** (LHC). The LHC is the world's largest and most powerful particle accelerator. Originally, large particle accelerators were termed colliders, and were designed to investigate the nature of matter by examining the structure of atoms and molecules via collisions. In contrast, a **synchrotron light** source is designed to use electrons to generate beams of infrared, UV, visible and X-ray radiation.

Synchrotron light, or synchrotron radiation, is the term given to a range of electromagnetic radiation of wavelengths from approximately 10^{-3} to 10^{-10} m.

A synchrotron is a doughnut-shaped particle accelerator designed to circulate electrons around a closed path at speeds very close to that of light. All particle accelerators require a source of charged particles. In a cathode ray tube these are provided by a device called an electron gun. Electrons are 'boiled' off a heated wire element and then accelerated from rest across a chamber emptied of air by an electric field created between the charged plates. The force acting on the charged particles from electric and magnetic fields will change the direction of the beam. As the charged particles are accelerated by these forces, work is done and the charges gain kinetic energy.

Cathode ray tubes are useful particle accelerators but are limited to using voltages over a few tens of kilovolts. A linear accelerator, or **linac**, accelerates particles in straight lines (shown in Figure 8.10). The Australian Synchrotron and electron linacs make use of travelling waves to accelerate particles. They consist of an electron gun, a vacuum system, focusing elements (elements that act on the beam to constrict it to a narrow beam) and RF (radio-frequency) cavities (shown in Figure 8.10). Electrons are emitted from an electron gun in pulses and are accelerated through the linac by powerful bursts of radio-frequency (RF) radiation. Electrons then travel around a **booster ring**, and are accelerated further as they pass through RF cavities until reaching an energy of 3 GeV. The magnetic field of the bending magnets is increased as the velocity of electrons increases within the booster ring.

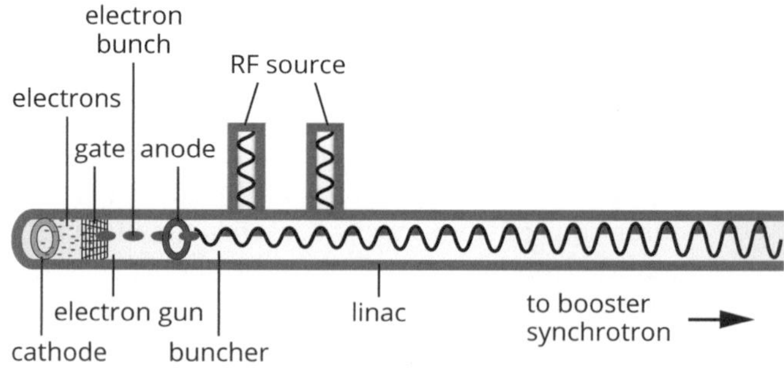

FIGURE 8.10 A diagram of a travelling-wave linac

Electrons are then channelled into the **storage ring**. The storage ring is a doughnut-shaped tube that keeps the electrons in the circular path by a series of bending magnets, known as dipole magnets, that make them bend in arcs as they orbit at speeds near that of light. Synchrotron radiation (light) is produced as the particles travel through the strong magnetic fields of the dipole magnets. Electrons replace the energy lost due to this production of synchrotron light when they pass through RF cavities in the storage ring. Synchrotron light leaves the storage ring, passing down **beamlines** to a number of independent experimental stations known as endstations.

ISBN 978 1 4886 1936 6

WORKSHEET 8.1

Knowledge review—the structure of matter

Check your current knowledge of the origins of the universe and atomic radiation by stating whether the statements below are true or false. For the statements you've marked as false, give a short explanation of the correct alternative.

Statement	T/F	Explanation
Beta radiation can be negatively charged.		
The atomic number of an element is the number of protons in a nucleus of that element.		
Light from the Sun takes about 10 s to reach the Earth.		
The half-life of an element is always measured in years.		
The mass number of an isotope is the number of neutrons in a nucleus of that element.		
Alpha radiation cannot penetrate far into the skin.		
Marie Curie invented the Geiger counter.		
In every half-life, 50% of the remaining atoms will decay.		
Every radioactive isotope will last for exactly two half-lives.		
The kinetic energy of a particle will double if the particle's speed is doubled.		
The nuclear reaction taking place in the Sun's core is fusion.		
Kinetic energy is always conserved in collisions.		
Nothing can travel faster than 99.9% of the speed of light.		
Gamma and beta radiation are both types of electromagnetic radiation.		
When an atom decays and produces an alpha particle, a new element is formed.		
The Earth was formed about 400 million years ago.		
Carbon dating only works on the remains of living organisms.		
The most powerful ionising radiation is alpha radiation.		
Smoke detectors commonly use a radioactive isotope.		
In every atom the number of protons equals the number of neutrons.		
Both atomic bombs and nuclear reactors can use fission reactions.		
The atomic symbol for carbon-12 is ${}^{12}_{7}C$.		
A magnetic field cannot increase the kinetic energy of a charged particle.		
Gamma rays have a longer wavelength than ultraviolet rays.		
Every nuclear fission reaction produces three neutrons.		

WORKSHEET 8.2

Spectra—explaining an expanding universe

By considering hydrogen's absorption line in the red part of the spectrum, astrophysicists can work out how much stars and galaxies are moving away from Earth. Three stars have been identified to be moving away from Earth (Star 1, Star 2 and Star 3) in comparison to the hydrogen spectrum.

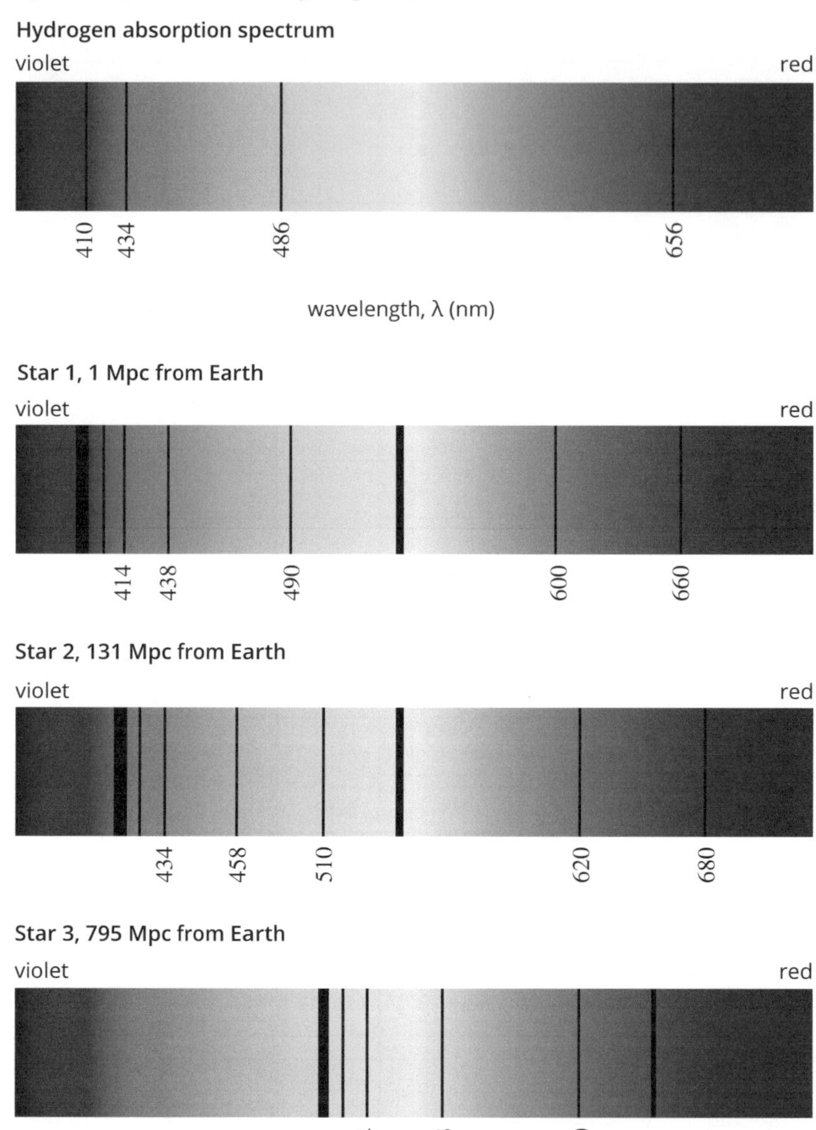

1 **a** For each of the stars above, find the cosmological redshift by using the formula below:

$$z = \frac{\lambda_{obs} - \lambda_{rest}}{\lambda_{rest}}$$

where z is the cosmological redshift, λ_{rest} is the red line position of hydrogen and λ_{obs} is the observed red line position of the star.

Star	Red line position (nm)	Amount of redshift (nm)	Cosmological redshift
1			
2			
3			

 ISBN 978 1 4886 1936 6

b Using the cosmological redshift found in part **a**, calculate the recessional speed (in $km\,s^{-1}$). This can be found by using the formula below, for small velocities:

$$z \approx \frac{v}{c}$$

Star	Distance (Mpc)	Recessional speed ($km\,s^{-1}$)
1	1	
2	131	
3	795	

c Draw a graph of distance versus recessional speed to find Hubble's constant.

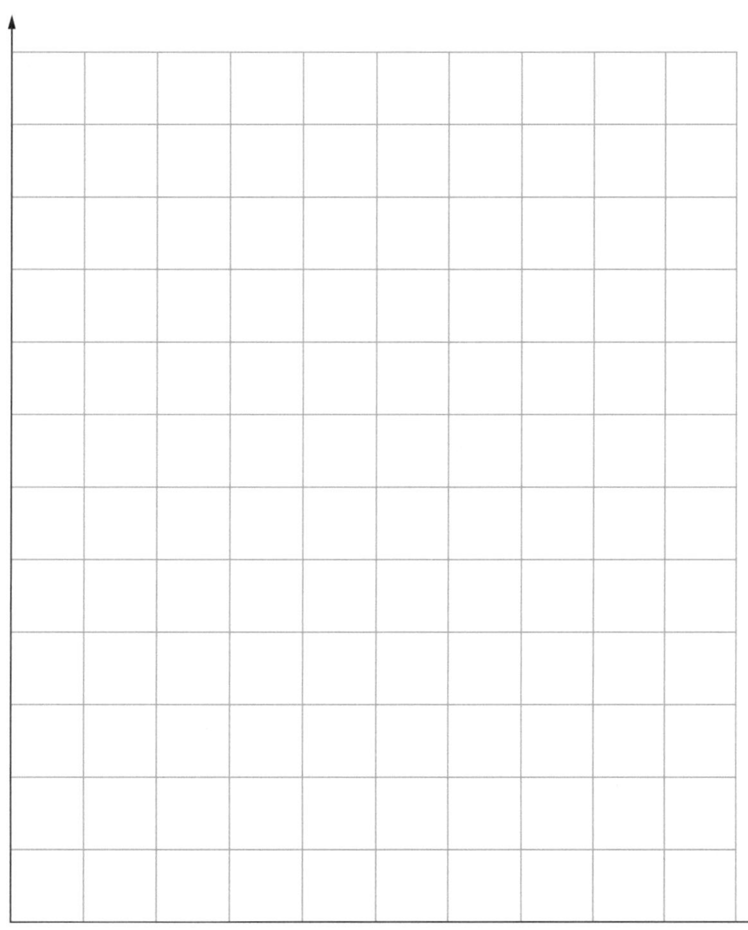

d Using this value for Hubble's constant, what is the approximate age of the universe?

e If the peak wavelengths emitted by each of the three stars are as shown in Table 1, rank the stars in order of hottest to coldest, where 1 is the hottest.

TABLE 1

Star	Peak wavelength ($\times 10^{19}$)	Ranking of surface temperature (hottest = 1)
1	5552	
2	2798	
3	7100	

WORKSHEET 8.2

2 Astrophysicists use spectroscopy to analyse light from distant stars. In the figure below, 16 stars have been identified according to their spectral class. It is your job as an astrophysicist to group the stars according to their spectral class. In order to do this you must find features common to a number of stars, and they must all fit into four categories.

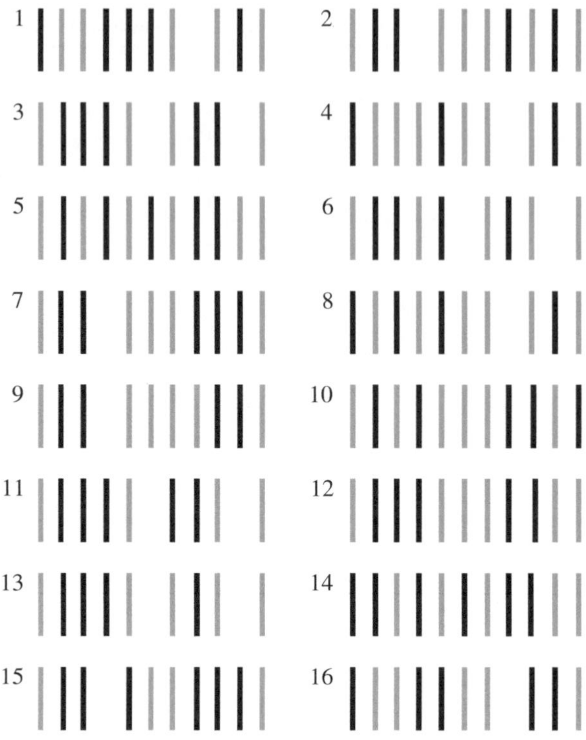

a

b

c

d

RATING MY LEARNING	My understanding improved	Not confident ←→ Very confident ○ ○ ○ ○ ○	I answered questions without help	Not confident ←→ Very confident ○ ○ ○ ○ ○	I corrected my errors without help	Not confident ←→ Very confident ○ ○ ○ ○ ○

 ISBN 978 1 4886 1936 6

WORKSHEET 8.3

The Hertzsprung–Russell (H–R) diagram

The Hertzsprung–Russell (H–R) diagram is a basic tool for astronomers. It shows some of the brightest stars visible from Earth. The figure below shows the common way of arranging the H–R diagram, with luminosity on the vertical axis (brighter at the top) and temperature on the horizontal axis. Each axis is labelled with two different scales. Temperature is derived from spectral type, so these are both plotted on the horizontal axes. Luminosity is derived from the absolute magnitude, so these are plotted together on the vertical axes. Note that the scales used in the H–R diagram for either luminosity or temperature are logarithmic, which means that equal distances along the horizontal and vertical axes do not show equal changes.

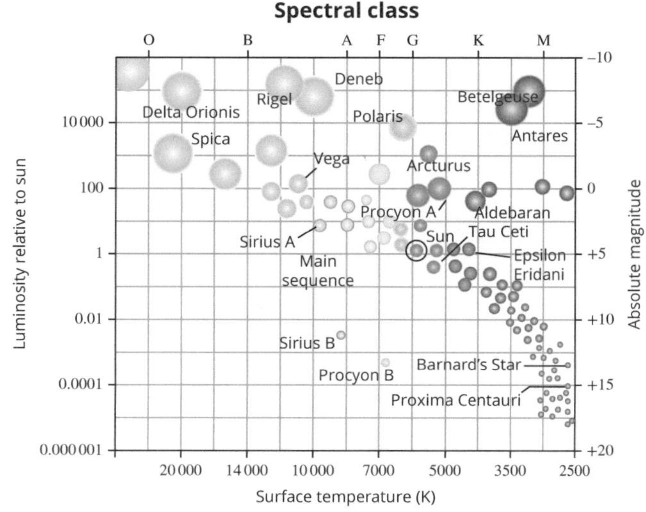

1 Explain the difference between white dwarfs, red giants and super giant stars. Show on the diagram where each would be found, based on the stars shown.

2 The star Orion (not plotted in the H–R diagram above) has a luminosity value of 30, and a surface temperature of about 8000°C. Vega, Sirius A and Orion are all about the same distance from the Earth. Name the correct order of these stars when ranked from brightest to dimmest.

3 How is it possible for white dwarf stars to have lower luminosity than the Sun even though the Sun is cooler than white dwarfs? Give an example of a white dwarf from the H–R diagram along with your explanation.

4 Using the H–R diagram, compare Polaris and Barnard's Star by referring to the percentage difference of surface temperature and approximately how much brighter one star is relative to the Sun.

5 Explain the difference between absolute magnitude and apparent magnitude.

6 The table below shows the absolute magnitude and surface temperature of some other stars.

Stars	Absolute magnitude	Surface temperature (K)
Capella	–0.7	4940
Achernar	–1.0	15000
Beta Centauri	–4.1	25500
Altair	+2.2	7500
Alpha Crucis	–4.0	26000
Pollux	+0.8	4865
Formalhaut	+2.0	8590
Deneb	–6.9	8525
Beta Crucis	–4.6	25000
Regulus	–0.6	10300
Epsilon Canis Majoris	–5.4	22200
Castor	–0.9	10300

a Draw a graph of this data with absolute magnitude on the *y*-axis and surface temperature on the *x*-axis.

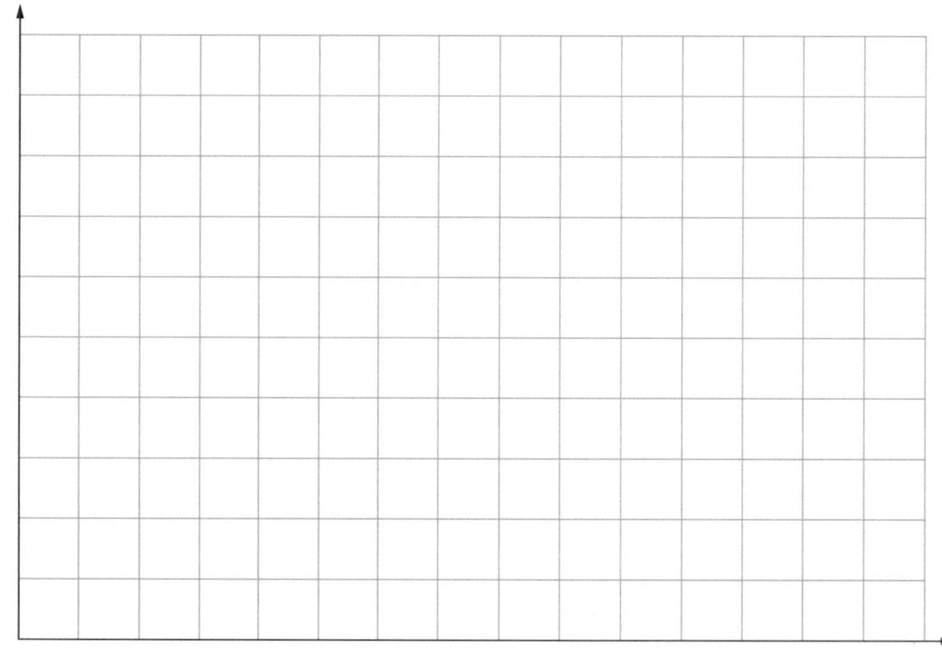

b Of the stars on your graph, which two have the highest positive numbers for their absolute magnitude? What is significant about these two stars?

c What is the most common spectral class?

RATING MY LEARNING	My understanding improved	Not confident ◄——► Very confident ○ ○ ○ ○ ○	I answered questions without help	Not confident ◄——► Very confident ○ ○ ○ ○ ○	I corrected my errors without help	Not confident ◄——► Very confident ○ ○ ○ ○ ○

 ISBN 978 1 4886 1936 6

WORKSHEET 8.4

Looking for particles

One of the earliest indications that atoms were not simply tiny hard spheres came when a high voltage was applied to a glass tube containing a very low-pressure gas. The gas started to emit light consisting of several discrete colours. Each gas emitted a different mix of wavelengths, indicating that there was a possible inner structure in the gas atoms. Further experiments were carried out using glass tubes containing virtually no gas at all. In these circumstances a green glow was observed in the tube wall near the positive electrode. It seemed as if there might be some type of emanation or radiation coming from the negative electrode—and thus cathode rays were discovered. We now know that the cathode rays are actually particles—electrons—being emitted from the negative electrode and accelerated towards the positive electrode.

1 If a potential difference of 6.0 kV is applied across an evacuated glass tube, how much energy is transferred to each electron travelling from one electrode to the other? (Use $e = 1.602 \times 10^{-19}$ C.)

2 How fast would the electron be travelling when it reaches the positive electrode? (Use $m_e = 9.109 \times 10^{-31}$ kg.)

It was soon determined that the 'rays' were deflected by an external electric field, in the opposite direction to the field, indicating that they were negatively charged. They were also deflected by a magnetic field. In 1897, J. J. Thomson was able to apply electric and magnetic fields to the rays and thereby find the charge-to-mass ratio (e/m) for the electron.

3 In the diagram, an electron enters a region where there is a magnetic field, B. In which direction will the magnetic force on the electron act?

4 If the force from the electric field created by the parallel plates acts in the opposite direction, which plate is positive?

5 When the fields are adjusted, the electrons can be made to travel straight through, undeflected. Show how the velocity of the undeflected electrons can be found.

6 Explain why this arrangement of fields is known as a 'velocity filter'.

The challenge was to now find the charge of the electron, as this would lead to its mass. In 1909, Robert Millikan's landmark experiment did just that. Millikan was able to observe microscopic oil droplets between a pair of charged plates. Many of the oil droplets were charged and were seen to move up or down at different speeds. They were so small that they reached a terminal velocity very quickly when acted upon by the forces of gravity, electrostatic attraction and air resistance.

When terminal velocity is reached, the net force is zero. The driving force, the sum of the oil droplet's weight and any electric force, is balanced by air resistance, and the size of the terminal velocity itself is proportional to the driving force.

7 Millikan found that the terminal velocities were clustered together at values that were multiples of each other. What did this clustering imply about the charge on the oil droplets?

8 Millikan then concentrated on the droplets that were moving very slowly. He adjusted the electric field between the plates until they were stationary. From the size of the droplet, he could calculate the droplet's mass. Consider an oil droplet of mass 1.72×10^{-16} kg held motionless between a pair of plates 2.50 mm apart at a potential difference of 90.0 V. How many elementary charges does the droplet carry?

From 1908 to 1913, around the same time as the American Millikan was performing his experiments, a pair of Ernest Rutherford's associates at the Cavendish Laboratory in England were using alpha particles to probe the atoms in gold foil. Hans Geiger and Ernest Marsden encased the alpha particle source in a lead shield with a small hole in it. The emerging narrow beam of alpha particles was directed at very thin gold foil merely hundreds of atoms thick. Most of the alpha particles passed straight through the gold foil as if it wasn't there. However, some underwent huge deflections, even emerging from the foil in the opposite direction.

9 Explain how this observation implied the existence of a very dense mass at the centre of the gold atom.

 ISBN 978 1 4886 1936 6

WORKSHEET 8.4

In 1932, Chadwick allowed the intense alpha particle radiation from polonium to strike a thin sample of beryllium.

10 The result was a beam of energetic radiation which consisted of particles that were unaffected by electric or magnetic fields. Which particles known at the time were ruled out by this observation?

11 The radiation, although very intense, could not be made to produce any photoelectric effect. This further ruled out which other particles?

It is now known that this nuclear reaction had the equation ${}^{4}_{2}\alpha + {}^{9}_{4}\text{Be} \rightarrow {}^{x}_{y}w + {}^{1}_{0}\text{n}$.

12 Find the values of w, x and y in the equation above.

13 A typical neutron produced in this reaction would be travelling at 10% of the speed of light. Given the neutron's mass of 1.675×10^{-27} kg, calculate the kinetic energy of such a neutron in units of MeV.

RATING MY LEARNING	My understanding improved	Not confident ⟵⟶ Very confident ○ ○ ○ ○ ○	I answered questions without help	Not confident ⟵⟶ Very confident ○ ○ ○ ○ ○	I corrected my errors without help	Not confident ⟵⟶ Very confident ○ ○ ○ ○ ○

WORKSHEET 8.5

Quantum mechanical nature of the atom

1 What was Rutherford's main contribution to our understanding of the atom? How did he determine this?

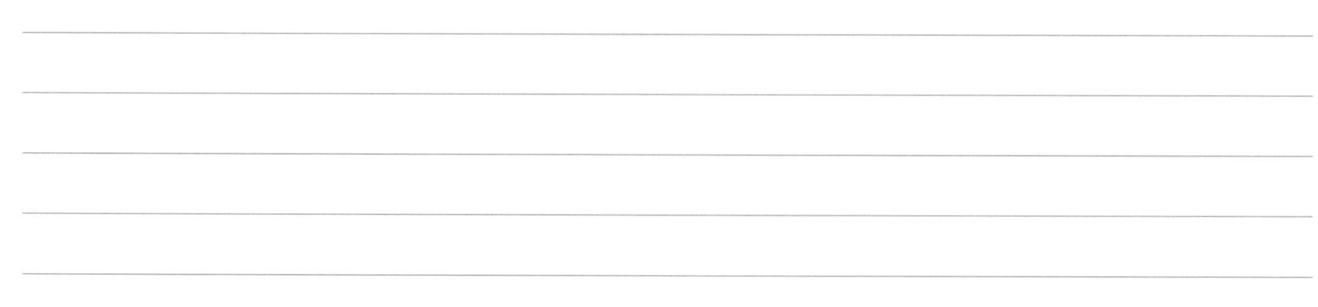

2 The absorption (a) and emission (b) spectra for hydrogen are shown below.

a Using the following diagram, which shows an electron at $n = 1$, explain the conditions under which absorption occurs.

$n = 3$

$n = 2$

$n = 1$

b Using the following diagram, which shows an electron at $n = 3$, explain how emission occurs. Draw all of the possible transitions.

$n = 3$

$n = 2$

$n = 1$

ISBN 978 1 4886 1936 6

WORKSHEET 8.5

3 Refer to the diagram below to answer the questions that follow.

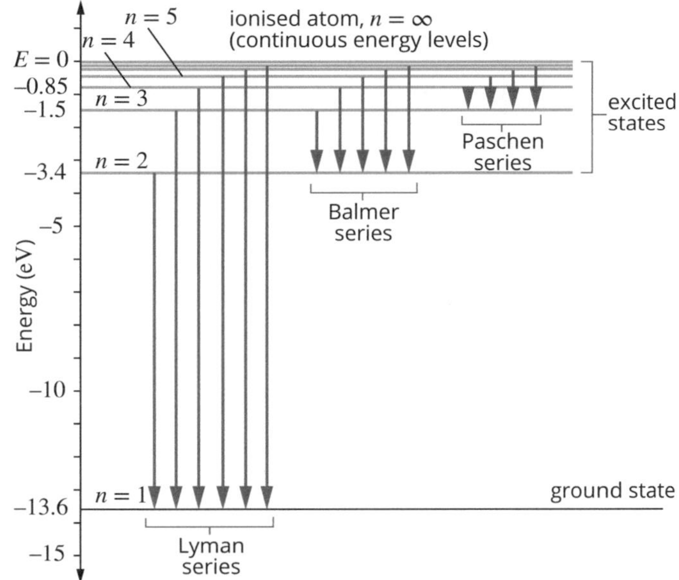

a Calculate the wavelength of a photon emitted when an electron transitions from the $n = 2$ to the $n = 1$ level.

b An already excited hydrogen atom undergoes an absorption transition from $n = 3$ to $n = 4$. Use the Rydberg expression $\frac{1}{\lambda} = R\left[\frac{1}{n_f^2} - \frac{1}{n_i^2}\right]$ to calculate the wavelength where this should be observed.

c For each of the following emission lines, identify the transition and whether it belongs to the Lyman series, the Balmer series or the Paschen series.

2.55 eV:

0.65 eV:

13.056 eV:

(Use the expression $E_n = -\frac{13.6}{n^2}$ for the last one.)

4 What were the limitations of the Bohr model?

5 Shown below are the energy levels for a mercury atom.

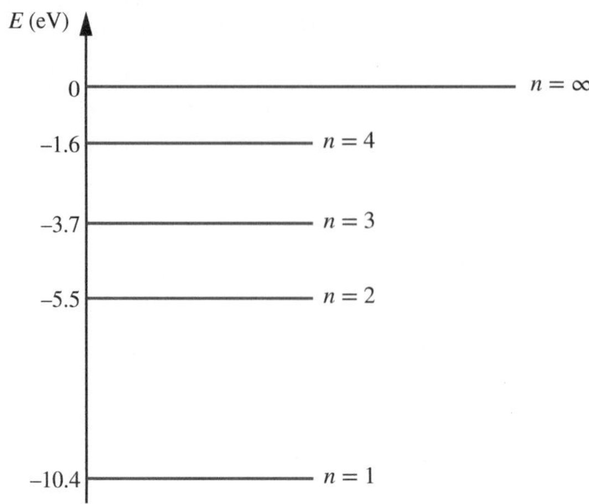

Light with photon energies of 5.5 eV, 8.8 eV and 9.6 eV is incident on mercury. Draw on the diagram all of the possible absorption transitions.

RATING MY LEARNING	My understanding improved	Not confident ◄——► Very confident ○ ○ ○ ○ ○	I answered questions without help	Not confident ◄——► Very confident ○ ○ ○ ○ ○	I corrected my errors without help	Not confident ◄——► Very confident ○ ○ ○ ○ ○

 ISBN 978 1 4886 1936 6

WORKSHEET 8.6

Matter waves

To answer the questions below, ignore relativistic effects in all calculations. Use $m_e = 9.1 \times 10^{-31}$ kg.

1 **a** How was Young's double-slit experiment modified to show matter acting like a wave?

b What other evidence was there for matter waves?

2 The de Broglie wavelength is given by $\lambda = \frac{h}{p}$ or $\lambda = \frac{h}{mv}$.

a An electron is fired with a velocity of 5.0×10^6 m s^{-1}. Calculate its de Broglie wavelength.

b Consider a 900 kg car travelling at 60 km h^{-1}. Calculate its de Broglie wavelength.

c What conclusions can you draw from the wavelengths in parts **a** and **b**?

3 **a** An electron is accelerated at 2.0 keV in an electron microscope. Calculate the de Broglie wavelength of the electron.

b How does this compare to the wavelength of visible light? Why is this useful in an electron microscope?

4 **a** Photon momentum has been used in solar sails for spaceflight. Calculate the momentum from one green-light photon with wavelength of 510 nm.

b Compare this to the momentum of a 100 g ball travelling at 15 km h^{-1}.

5 How did de Broglie modify the Bohr model using matter waves? Draw a diagram to assist your explanation.

6 How did Schrödinger further modify the Bohr model?

RATING MY LEARNING	My understanding improved	Not confident ◄──► Very confident ○ ○ ○ ○ ○	I answered questions without help	Not confident ◄──► Very confident ○ ○ ○ ○ ○	I corrected my errors without help	Not confident ◄──► Very confident ○ ○ ○ ○ ○

 ISBN 978 1 4886 1936 6

WORKSHEET 8.7

Half-life in radioactive decay

1 Thorium 231 decays to protactinium-231 with a half-life of 25 hours. For a sample with 5×10^{22} nuclei of thorium-231, plot the number of thorium-231 nuclei against time from 0 hours to 150 hours.

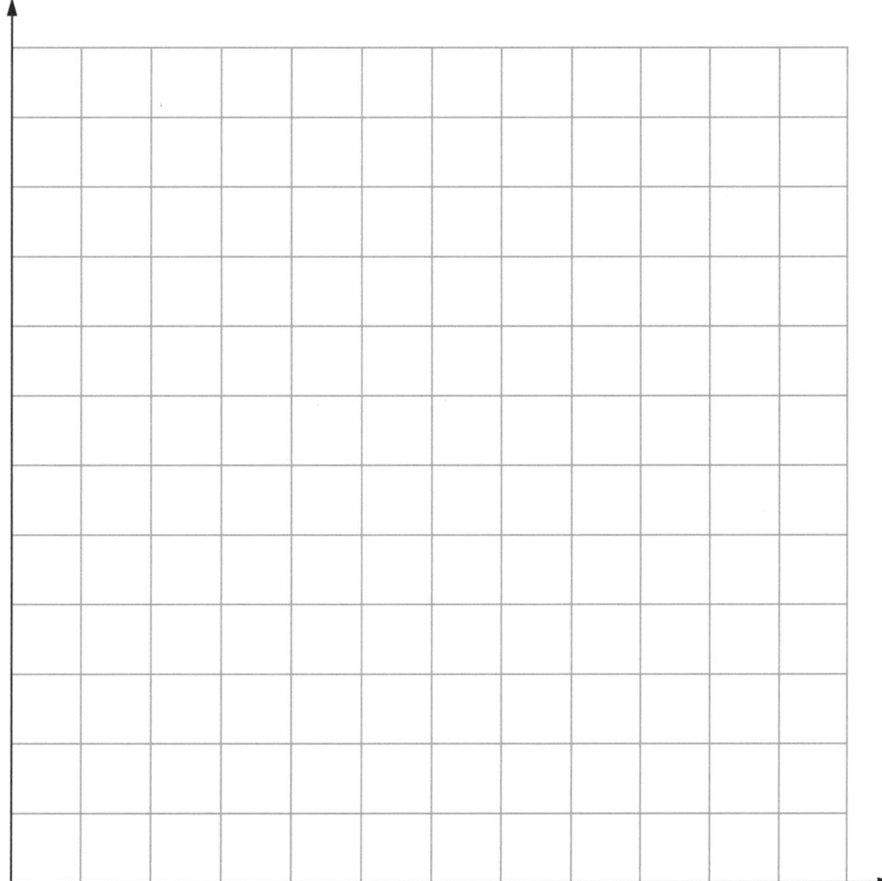

From the graph, what would be the expected number of atoms at:

a 30 hours? ____________________

b 70 hours? ____________________

2 ^{215}Po decays to ^{215}At with a half-life $t_{1/2} = 3.9\,s$. If a sample with 5×10^{22} nuclei starts off with 100% ^{215}Po, calculate the number of ^{215}Po at:

a $t = 7.8\,s$ ____________________

b $t = 15.6\,s$ ____________________

c $t = 10\,s$ ____________________

d $t = 18\,s$ ____________________

WORKSHEET 8.7

3 a A sample of ^{208}Tl measures 1200 counts in 5 s. What is its activity in Bq?

b What is its expected activity at 6.2 min given its half-life is 3.1 min?

c What is its expected activity at 8 min?

d What is its expected activity at 11 min?

4 A sample measures an activity of 3200 Bq. Ten minutes later it measures 200 Bq. What is the half-life of the sample in minutes?

5 A sample has an activity of 3000 Bq. Three days later it has an activity of 400 Bq. Calculate the half-life in hours.

RATING MY LEARNING	My understanding improved	Not confident ◄──► Very confident ○ ○ ○ ○ ○	I answered questions without help	Not confident ◄──► Very confident ○ ○ ○ ○ ○	I corrected my errors without help	Not confident ◄──► Very confident ○ ○ ○ ○ ○

 ISBN 978 1 4886 1936 6

WORKSHEET 8.8

Nuclear energy

1 atomic mass unit (u) = 1.66054×10^{-27} kg

Particle	Mass (u)
proton	1.0072766
neutron	1.0086654
electron	0.000548

1 Consider a carbon-12 nucleus, which has a mass of exactly 12.00 u.

a Calculate the mass of all the individual nucleons in the carbon-12 nucleus.

b Calculate the mass defect, Δm, between the mass of the individual nucleons and that of the carbon-12 nucleus.

c Where does this difference in energy go?

d Calculate this energy in J and MeV using Einstein's famous equation.

e One way of determining the stability of a nucleus is to look at the binding energy per nucleon. Calculate the binding energy per nucleon for carbon in MeV per nucleon.

2

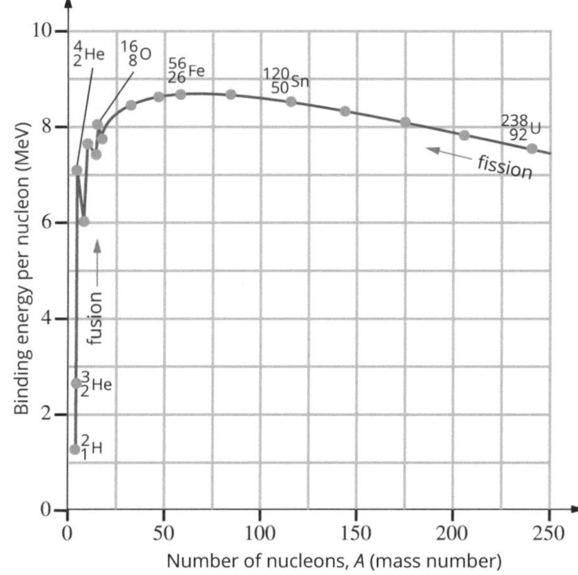

ISBN 978 1 4886 1936 6

a Calculate the binding energy per nucleon (in MeV) for the isotope $^{3}_{2}He$, given that its mass is 3.016029 *u*.

b Calculate the binding energy per nucleon (in MeV) for the isotope $^{235}_{92}U$, given that its mass is 235.043923 *u*.

Based on the graph above:

c Which is the most stable nucleus?

d Which elements are most suitable for fusion, and why?

e Which elements are most suitable for fission, and why?

3 By comparing the masses of the reactants and the products, calculate the energy released in the decay of bromine-82 (81.916204 *u*) to krypton-82 (81.913484 *u*).

RATING MY LEARNING	My understanding improved	Not confident ◄——► Very confident ○ ○ ○ ○ ○	I answered questions without help	Not confident ◄——► Very confident ○ ○ ○ ○ ○	I corrected my errors without help	Not confident ◄——► Very confident ○ ○ ○ ○ ○

 ISBN 978 1 4886 1936 6

WORKSHEET 8.9

Fission versus fusion

1 atomic mass unit (u) = 1.66054×10^{-27} kg

Particle	Mass (u)
proton	1.0072766
neutron	1.0086654
electron	0.000548

1 Nuclear fission is produced by firing a neutron at uranium-235 (235.0439 u). In this case, the fission products are rubidium-93 (92.922 u) and cesium-140 (139.917 u).

a Write a balanced nuclear equation for this reaction and hence determine how many neutrons are released.

b Why is a neutron used instead of an alpha or beta particle?

c Compare the energy of the incident neutron with the released neutrons.

d Explain what needs to happen in order for the released neutrons to cause fission in other uranium nuclei.

2 **a** Natural uranium is often found with uranium-238 and uranium-235. State the percentage of uranium-235 found in the uranium, and the process needed in order to use the uranium in a nuclear reactor.

b What is the difference between a controlled nuclear chain reaction, and an uncontrolled chain reaction that occurs in an explosion?

c Explain the measures taken to control nuclear reactions in a reactor.

3 **a** Calculate the energy released in the reaction in question 1.

b If one mole of uranium-235 (i.e. 235 g) undergoes fission, what is the total power output?
1 mol = 6.02×10^{23} molecules

4 During a reaction, 73% of plutonium-239 (239.052163 *u*) undergoes fission and releases fragments of xenon-134 (133.905395 *u*) and zirconium-103 (102.92660 *u*).

a Calculate the energy released by each nucleus in J and MeV.

b Calculate the energy released by 239 g of plutonium in J.

5 In stars, deuterium (2.014102 *u*) and hydrogen (1.007825 *u*) can fuse together to form helium-3 (3.016029 *u*) with the release of a gamma ray.

a Write a balanced nuclear reaction for this.

b Calculate the energy released in this reaction for one deuterium and one hydrogen nucleus.

6 **a** Scientists have been trying create fusion reactors on Earth. Why is this a difficult process?

b What advantages would a fusion reactor have over fission reactors?

RATING MY LEARNING	My understanding improved	Not confident ◄──► Very confident ○ ○ ○ ○ ○	I answered questions without help	Not confident ◄──► Very confident ○ ○ ○ ○ ○	I corrected my errors without help	Not confident ◄──► Very confident ○ ○ ○ ○ ○

 ISBN 978 1 4886 1936 6

WORKSHEET 8.10

Deep inside the atom

PURPOSE

Your task is to find out more about subatomic physics. Below is a list of questions or topics for you to tackle. In each case, you need to do two things.

1 Find out some more information, beyond what is available in your text book.

2 Produce a brief answer or explanation of the topic.

It is suggested that you look at four of these questions, although time constraints may change this. You should produce no more than a paragraph or two for each one.

QUESTIONS

1 Outside a nucleus, free neutrons decay. What is the decay half-life and equation? How do the masses of the neutron and its decay products help explain the process?

2 What is the process of positron decay? When and where does it happen? What about the masses and energy involved?

3 How do protons stick together inside the nucleus when they are all positively charged? What do neutrons have to do with this? Where does the 'liquid-drop' model fit in?

4 Where and when were mesons discovered? How do they decay? Consider their masses and charges.

5 There are four fundamental forces in nature, but, although useful, the idea of 'force fields' is no longer current. Even Newton was uncomfortable with 'action-at-distance'. How are forces mediated by particles? What are the particles involved for each force?

6 Particle accelerators have steadily increased in power, now being able to generate thousands of new particles not 'seen' since the big bang. How is this done? How can heavy particles be created out of nothing?

7 Briefly explain the terms hadron, lepton, boson, baryon and fermion. Where do quarks fit into this model?

8 If there are quarks inside all nuclei, why have free quarks not been directly observed?

9 Where does antimatter fit into the Standard Model? What happens when matter meets antimatter? Why does the universe seem to consist only of matter and not significant amounts of antimatter?

REQUIREMENTS AND CONSTRAINTS

In each case, focus on the conservation laws of physics:

- conservation of momentum
- conservation of angular momentum
- conservation of charge
- conservation of mass–energy.

There are also other conservation laws which apply. They may also be relevant.

Do not simply copy what you have found. Put your explanation into your own language and use only the concepts you have covered in your course.

COMMUNICATING

Before committing yourself to the final version of your answers, talk over what you have found with other students in your class. Check if your description makes sense to them.

RATING MY LEARNING	My understanding improved	Not confident ◄──► Very confident ○ ○ ○ ○ ○	I answered questions without help	Not confident ◄──► Very confident ○ ○ ○ ○ ○	I corrected my errors without help	Not confident ◄──► Very confident ○ ○ ○ ○ ○

WORKSHEET 8.11

Particle accelerators

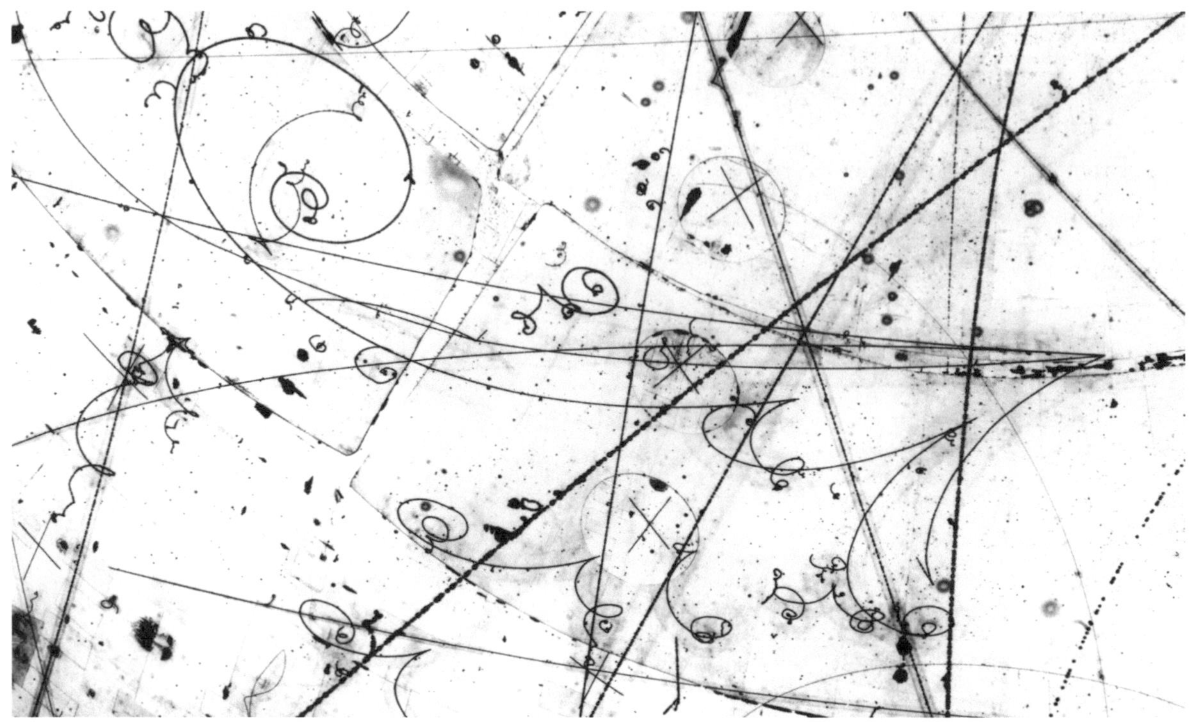

FIGURE 8.11 Photograph of a neutrino particle interaction in the 5 m bubble chamber of the Fermi National Accelerator Laboratory (Fermilab) near Chicago, USA

The simplest particle accelerator is arguably a Van de Graaff generator. This consists of a moving belt that carries charge to an insulated hollow metal sphere. Even simple versions of this device can produce very high voltages—up to 400 keV in the right conditions.

The actual process whereby positive charge accumulates on the upper sphere is a little complicated; essentially, charge is carried up the belt and removed at the top roller. Even though the metal sphere at the top gets more and more positively charged, the charges inside the roller are unaffected by the repulsion as they are effectively inside a Faraday cage.

The high voltages created can be used to accelerate charged particles inside a separate vacuum tube.

1 If the belt rotates at a frequency of 6.0 Hz, and 5.0 μC of charge is deposited on the belt every time it goes around, what is the effective current moving up to the top?

2 The potential difference can be doubled by putting two generators back-to-back, one being positively charged, the other negatively charged. How much energy will be given to 7.0 μC of charge in such a tandem 4.0×10^6 V Van de Graaff generator when it accelerates from the positive to the negative end?

Until the late 1920s, the Van de Graaff generator was the only way to effectively accelerate charged particles. The next development was the linear accelerator or linac. It consists of a series of hollow tubes of increasing length, connected alternately to positive and negative potentials as shown in the diagram above right.

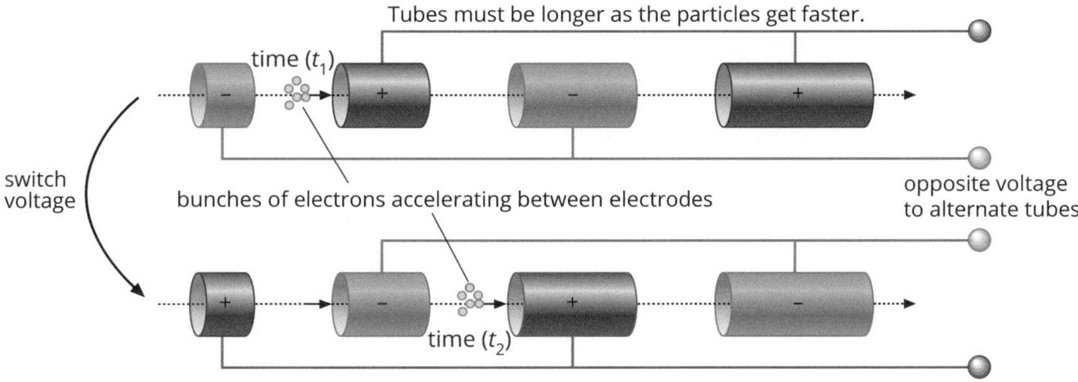

Linac diagram

In the initial acceleration phase, negative ions are injected into the linac from a source at ground potential (shown in the bottom diagram). While the ions are inside the first hollow tube, the charge on the tube is reversed. This does not affect the ions as the hollow tube constitutes a Faraday cage, so they continue at a constant speed. When the ions emerge from the tube they again are subjected to an accelerating potential across the gap to the next tube. This process is repeated until the ions eventually emerge from the last tube.

3 The alternating potential supplied to the tubes is created by a radio frequency (RF) source, typically at 12 MHz. If the potential difference at each stage is 800 V, what will be the effective potential difference applied to the ions in 500 μs?

4 The tubes increase in length as the particles get faster and faster, but near the end of a long linac they are nearly constant in length. Explain why this would be the case.

The main problem with a linac is that they need to be very long, which poses difficulties for construction and power supply. In 1932 Ernest Lawrence had the inspiration to wrap the operation of the linac into a smaller device by turning the hollow tubes into a pair of hollow semi-cylinders, known as 'dees', facing each other across a small gap, an arrangement he called the cyclotron. The ions would be forced into circular paths by placing the cylinders inside a strong magnetic field. The hollow cylinders act as a Faraday cage and have their charge quickly reversed while the ions are inside, so every time the ions emerge from a dee, they are further accelerated.

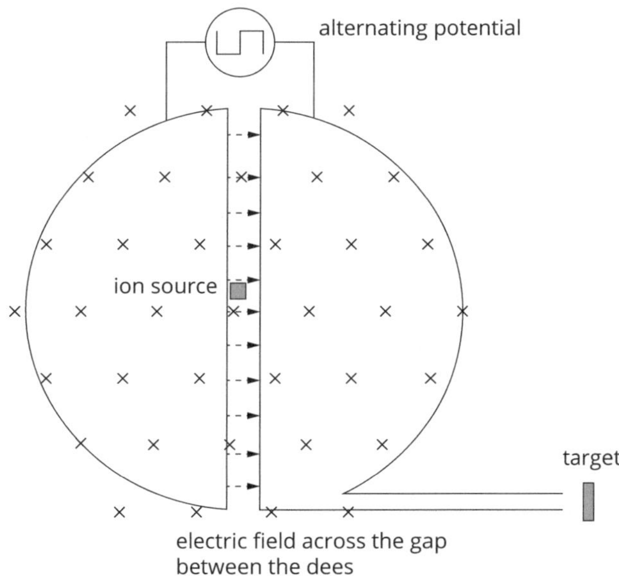

WORKSHEET 8.11

Consider an ion of mass 5.90×10^{-26} kg and charge 3.20×10^{-19} C, which is accelerated in the cyclotron and leaves from the maximum radius of 0.32 m to hit a target at the end of a tube. The magnetic field of the cyclotron has a magnitude of 1.20 T.

5 What is the sign of the charge on the ion?

6 What is the velocity of the ion as it hits the target?

7 What is its kinetic energy?

8 The electric field is depicted at just one instant. It is reversed at regular intervals by the alternating 4.0 V potential applied to the dees. How much energy is given to the ion each time it accelerates across the gap?

9 How many circuits does the ion make by the time it emerges from the cyclotron?

10 The final energy of the ions could be increased by having a stronger magnetic field and/or a bigger radius. Explain why increasing the size of the accelerating potential will have no effect.

RATING MY LEARNING	My understanding improved	Not confident ◄──► Very confident ○ ○ ○ ○ ○	I answered questions without help	Not confident ◄──► Very confident ○ ○ ○ ○ ○	I corrected my errors without help	Not confident ◄──► Very confident ○ ○ ○ ○ ○

 ISBN 978 1 4886 1936 6

WORKSHEET 8.12

Literacy review—the terminology of elements and matter

The study areas of nuclear and quantum mechanics introduce many new terms, a number of which sound quite similar. Complete the following table with the definition of each term, taking care to clearly distinguish each term from similar-sounding terms. You'll find these grouped together rather than being listed purely alphabetically.

Term	Definition
alpha particle	
beta particle	
gamma ray	
big bang	
beamline	
booster ring	
cyclotron	
decay series	
electron gun	
emission spectrum	
energy levels of atoms	
endstation	
fissile	
fission	
fusion	
half-life	
Hertzsprung–Russell (H–R) diagram	
atomic number	
mass number	
neucleons	
neutron	
neutrino	
antineutrino	
nuclide	

RATING MY LEARNING	My understanding improved	Not confident ◄——► Very confident ○ ○ ○ ○ ○	I answered questions without help	Not confident ◄——► Very confident ○ ○ ○ ○ ○	I corrected my errors without help	Not confident ◄——► Very confident ○ ○ ○ ○ ○

WORKSHEET 8.13

Thinking about my learning

On completion of Module 8: From the universe to the atom, you should be able to describe, explain and apply the relevant scientific ideas. You should also be able to interpret, analyse and evaluate data.

1 The table lists the key knowledge covered in this module. Read each and reflect on how well you understand each concept. Rate your learning by shading the circle that corresponds to your level of understanding for each concept. It may be helpful to use colour as a visual representation. For example:

- green—very confident
- orange—in the middle
- red—starting to develop.

Concept focus	**Rate my learning**				
	Starting to develop ◄——► Very confident				
The big bang theory and the expansion of the universe	○	○	○	○	○
Using stellar spectra to analyse properties of stars	○	○	○	○	○
Thomson's charge-to-mass experiment	○	○	○	○	○
Millikan's oil-drop experiment	○	○	○	○	○
The development of different models of the atom	○	○	○	○	○
de Broglie's matter waves	○	○	○	○	○
Alpha, beta and gamma radiation	○	○	○	○	○
Half-life	○	○	○	○	○
Fission versus fusion reactions	○	○	○	○	○
The Standard Model	○	○	○	○	○
Role of particle accelerators in modern physics	○	○	○	○	○

2 Consider points you have shaded from starting to develop to middle-level understanding. List specific ideas you can identify that were challenging.

3 Write down two different strategies that you will apply to help further your understanding of these ideas.

ISBN 978 1 4886 1936 6

PRACTICAL ACTIVITY 8.1

Detecting radiation with a G-M tube

Suggested duration: 40 minutes

INTRODUCTION

This is an introductory activity. Besides the obvious fact that none of our senses can detect individual decay events, the nuclear decay process seems to be random yet at the same time predictable. It is impossible to say which nucleus will become unstable enough to decay next; however, it is fairly easy to use a Geiger–Müller (G–M) tube to count the number of nuclei which do decay per second at all the locations in a radioactive sample. If you recorded the nuclear decay of a radioactive sample with a G–M tube and counter (or nuclear sensor) and plotted counts per time interval over a period of time, the results would look like a standard distribution curve.

Because in many ways nuclear radiation behaves as though the radiation were tiny 'bullets', it makes sense that different materials absorb the energy of nuclear radiation in different ways. The nature of the material through which nuclear radiation moves influences how much energy is absorbed.

MATERIALS

- Geiger–Müller tube, counter and stopwatch or data-collection system
- power supply and counter or computer interface
- base and support rod
- radioactive sources (alpha, beta, gamma)
- right-angle clamp
- shielding material: lead, paper, plastic and aluminium in 5 cm squares

PURPOSE

To investigate the penetrating ability of three common types of nuclear radiation and the ability of different materials to absorb the energy associated with nuclear radiation.

PROCEDURE

In the first part of this activity, the background radiation is measured. In the second part of the activity, the radiation from three different sources that are shielded with different thicknesses of three different materials is compared. The measurements are made from a fixed distance over equal intervals of time.

Connect the G–M tube to its power supply and counter supplied with the equipment or connect the G–M tube to your data collection system according to the manufacturer's instructions. Clamp the G–M tube securely in a retort stand, pointing downwards so the distance from each source is controlled.

The radioactive sources used in this experiment may be dangerous if not handled properly. Handled as follows, laboratory samples provide little or no cause for concern.

- Do not touch the sources with bare hands.
- Pick up sources with forceps or tongs.
- Do not point sources towards human bodies.
- Maintain a distance of about half a metre from samples when possible.
- Wash hands thoroughly after performing an experiment.
- Importation of new materials is now strictly controlled. Contact suppliers to ascertain availability.

PART A: Background radiation

1 Move all radiation sources at least 3 m from the G–M tube. Turn on the power supply and counter. Zero the counter, and then start the counter or start data collection with your data-collection system.

2 Allow the counter to record for 60 s and record the number of counts. Repeat three or four times during the course of the total activity to enable a reliable average to be found for the course of the activity. Make sure the samples are put away from the G–M tube or shielded each time a background check is taken.

3 Record the background counts in Table 1 in the Data and analysis section.

PART B: Radiation shielding

Use the same equipment set up as in Part A. The only difference in Part B is that you will be placing different radiation sources and different shielding materials directly below the G–M tube.

1 Position the sealed alpha source under the G–M without any shielding. Turn on the power supply and counter, if they are not on already, or connect your data-analysis system. Measure and record the number of counts over a 60 s period without any shielding over the alpha source. Record this in a spreadsheet set up like Table 2 of the Data and analysis section.

2 Now measure the radiation counts from the alpha source when it is shielded by one layer of paper. Repeat for two, three, four and five layers of paper. In each case, record the count for the 60 s sampling period.

3 Replace the sealed alpha source with a sealed beta source. Repeat the steps, recording counts for the unshielded beta source, and then repeat for the beta source with one to five layers of paper on top. Record the number of radiation counts per 60 s period for each trial.

4 Replace the sealed beta source with a sealed gamma source. Repeat the data-collection process again with no shielding and then with one to five sheets of paper.

5 Repeat steps 1–4 for each source using thin squares of plastic as the shielding material. Record the number of radiation counts per 60 s period for each trial.

6 Repeat the entire process of data collection for each source using thin squares of lead metal as the shielding material. Record the number of radiation counts per 60 s period for each trial.

7 Repeat the entire process of data collection for each source using thin squares of aluminium as the shielding material. Record the number of radiation counts per 60 s period for each trial.

DATA AND ANALYSIS

TABLE 1 Background radiation count

Sample	Counts
1	
2	
3	
4	
Average count:	

TABLE 2 Radiation shielding

Shielding material	Source	Counts per 60 s

 ISBN 978 1 4886 1936 6

PRACTICAL ACTIVITY 8.1

Construct a graph using the number of layers of shielding versus counts for each type of radiation and material. Graph all tests for each type of material clearly on one set of axes so that clear comparisons between sources can be made.

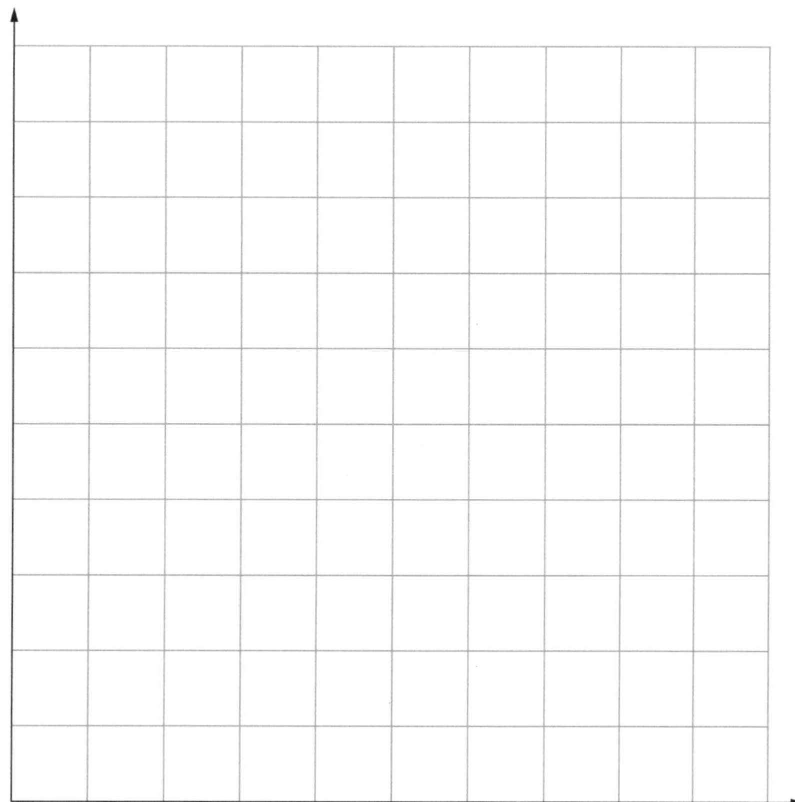

PRACTICAL ACTIVITY 8.1

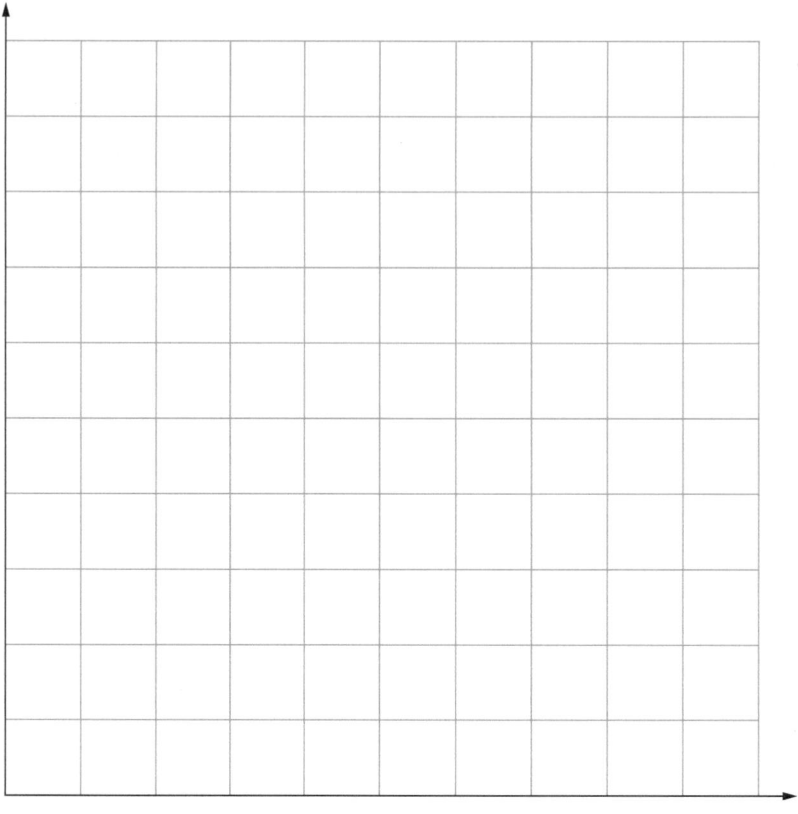

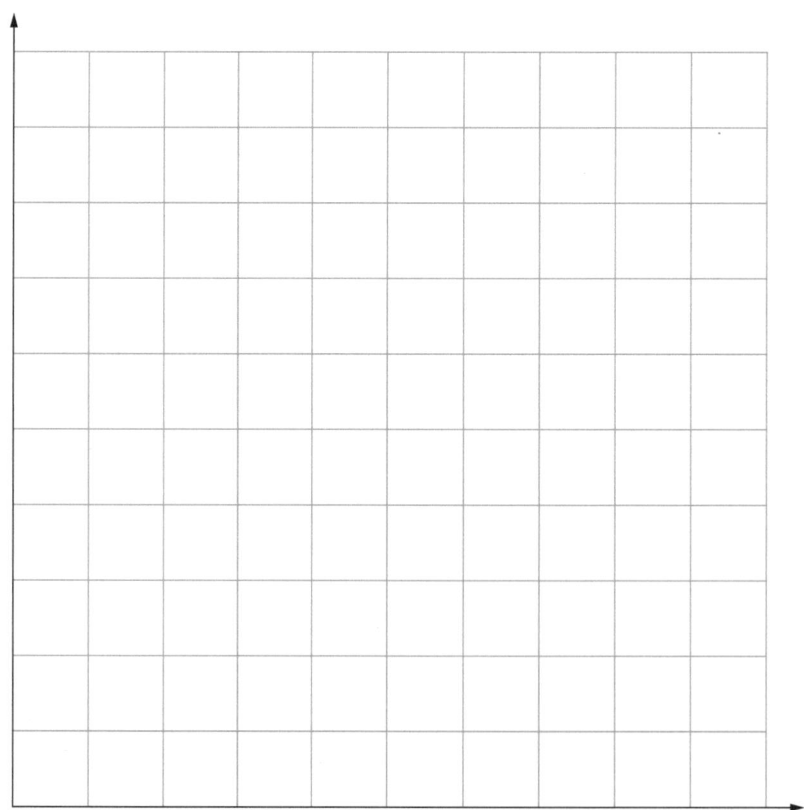

ISBN 978 1 4886 1936 6

PRACTICAL ACTIVITY 8.1

CONCLUSION

1 Summarise your results, commenting on the penetration of each source and the effect of shielding.

2 What generalisation can you make about the effect of the thickness of the shielding material on the count rate?

3 What generalisation can you make about the effect of the density of the shielding material on the count rate?

4 Why was background radiation measured? Do you expect the value to be constant?

RATING MY LEARNING	My understanding improved	Not confident ◄——► Very confident ○ ○ ○ ○ ○	I answered questions without help	Not confident ◄——► Very confident ○ ○ ○ ○ ○	I corrected my errors without help	Not confident ◄——► Very confident ○ ○ ○ ○ ○

PRACTICAL ACTIVITY 8.2

An analogue model of radioactive decay

Suggested duration: 20 minutes

INTRODUCTION

In this activity you will work as a whole class, recording results as a group and noting them down for your own analysis at the end of the practical component.

An early introduction to the process of radioactive decay, this activity often has its best results when it precedes the theory of radioactive decay. Used effectively, it is a means of introducing the randomness of decay and generating data upon which theory can effectively be developed.

MATERIALS

- class of students
- one coin per student
- one standard six-sided dice per student (optional as time allows)

PURPOSE

To model the randomness of radioactive decay.

PROCEDURE

1 All of the students in the class stand up. You represent the radioactive nuclei within a sample.

2 On the word 'go' or 'decay' from the teacher or group leader, start shaking a coin in your cupped hands.

3 Every 10 s, the teacher will give a signal. Stop shaking and look at the coin. If your coin shows 'heads', sit down. You represent a disintegrated nuclei.

4 Count the number of students that remain standing each time, and record the value on a whiteboard or other group recording device for later transfer to Table 1 in the Data and analysis section.

5 Continue until all students have 'decayed'.

6 Repeat two or three times as total lesson time permits.

7 Repeat with each student shaking a dice. Sit down if the dice shows the number 1 facing upwards.

DATA AND ANALYSIS

TABLE 1 Decay rates for a coin

Time (s)	Standing students, Trial 1	Standing students, Trial 2	Standing students, Trial 3
0			
10			
20			
30			
40			
50			
60			
70			
80			
90			
100			

ISBN 978 1 4886 1936 6

PRACTICAL ACTIVITY 8.2

Graph the results for each trial on the graph space provided. Show all trials on the one graph.

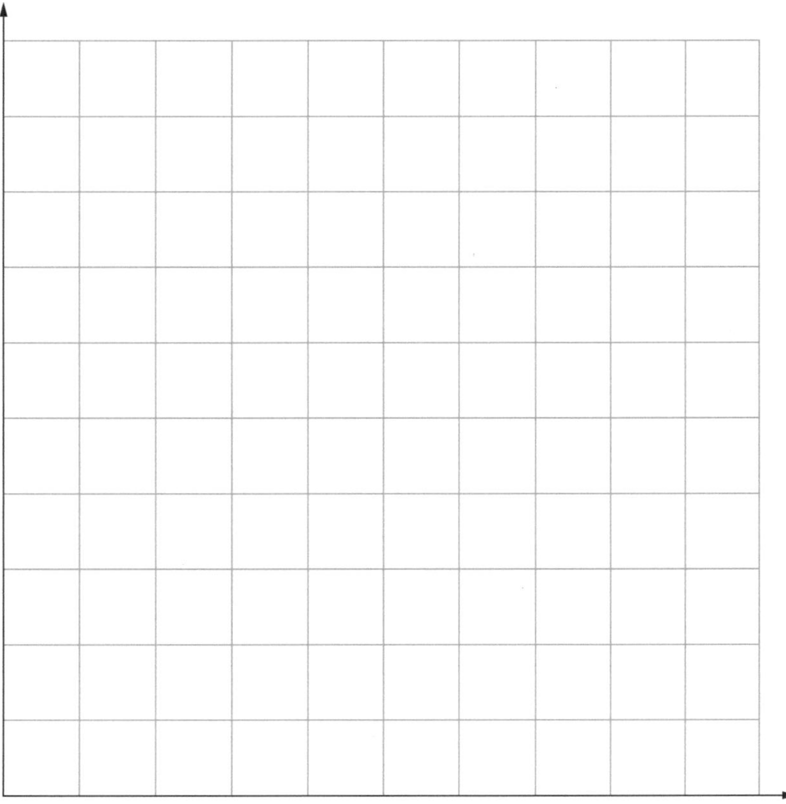

TABLE 2 Decay rates for a dice

Time (s)	Standing students, Trial 1	Standing students, Trial 2	Standing students, Trial 3
0			
10			
20			
30			
40			
50			
60			
70			
80			
90			
100			

PRACTICAL ACTIVITY 8.2

Graph the results for each trial on the graph space provided. Show all trials on the one graph.

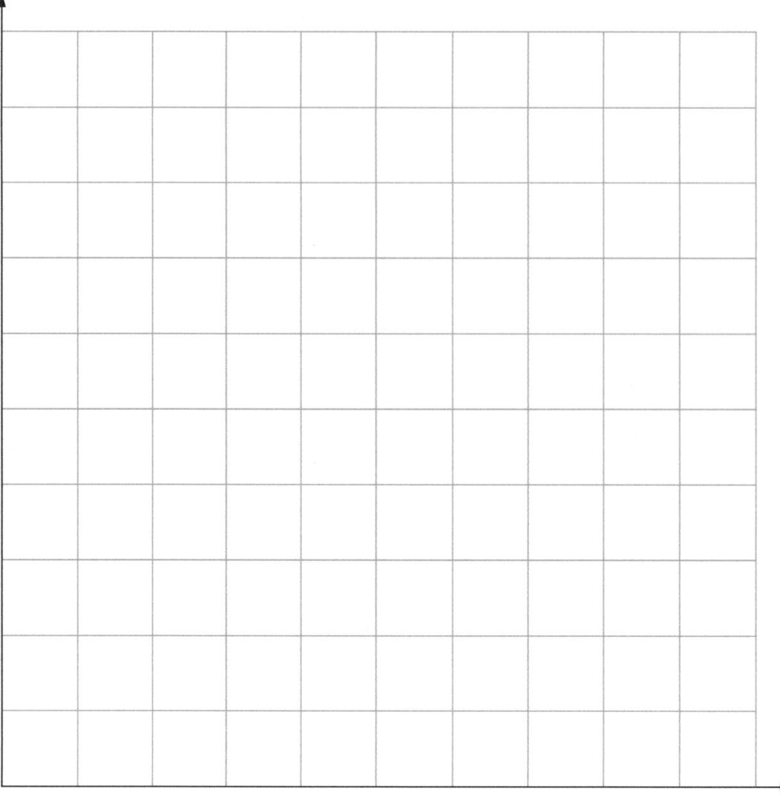

CONCLUSION

1 Comment on the relative decay rates for coin and dice. Was the rate of decay the same for each trial?

2 Is this process of sitting down a purely random process? Explain.

3 How does the number of 'radioactive nuclei' remaining decrease with time? Does it 'fit' any standard graphical form covered in maths? Suggest suitable models.

4 'The time for half of the "radioactive nuclei" to disintegrate is constant.' From the data obtained, do you think that this statement is true or false for this experiment? Explain and discuss your answer.

 ISBN 978 1 4886 1936 6

PRACTICAL ACTIVITY 8.2

5 If 1000 students were to take part in this experiment instead of one class, do you think the experimental result would agree more closely with the statement in question 4? Explain and discuss your answer.

6 When the statement in question 4 is true, the time for half of the nuclei to disintegrate may be called the half-life. What do you expect the half-life of the process described in this experiment to be? Calculate the 'half-life' of the class.

RATING MY LEARNING	My understanding improved	Not confident ◄——► Very confident ○ ○ ○ ○ ○	I answered questions without help	Not confident ◄——► Very confident ○ ○ ○ ○ ○	I corrected my errors without help	Not confident ◄——► Very confident ○ ○ ○ ○ ○

PRACTICAL ACTIVITY 8.3

Determining the half-life of an isotope

Suggested duration: 40 minutes

INTRODUCTION

In this activity, the radioactive decay of a short-lived isotope will be investigated. Radioactive (or nuclear) decay is a random process, yet it is somewhat predictable.

PURPOSE

To investigate the radioactive decay of an isotope and hence determine its half-life.

BACKGROUND

The number of decaying radioactive atoms within a period of time is proportional to the amount of radioactive atoms present. That is,

$$\frac{\Delta N}{\Delta t} = -\lambda N$$

where:

ΔN is the number of radioactive atoms that decays within the set time Δt

λ is the decay constant (the fraction of the radioactive atoms that decays per unit time) and is different for each isotope

N is the number of radioactive atoms present.

Mathematically, this equation takes the following form as a function of time:

$$N_t = N_0 e^{-\lambda t}$$

where:

N_0 is the number of radioactive atoms at time $t = 0$

N_t is the number of radioactive atoms after time t

λ is the decay constant for the isotope.

The half-life, $t_{1/2}$, of a radioactive isotope is the time it takes for half of the original atoms to decay.

$$t_{1/2} = \frac{\ln 2}{\lambda}$$

Half-lives can be as short as a fraction of a second or as long as a billion years, depending on the isotope.

MATERIALS

- G–M tube with counter and stopwatch or data-collection system
- isotope generator producing barium-137m solution

PROCEDURE

In this experiment, your teacher will use an isotope generator to provide you with a small quantity of the short-lived barium-137m isotope. The barium-137m isotope is a product of the decay of the cesium-137 isotope. The decay series is shown below.

^{137}Cs ($t_{1/2} = 30.1$ y)
- β 85% → ^{137m}Ba ($t_{1/2} = 2.55$ min) → γ → ^{137}Ba (stable)
- β 15% → ^{137}Ba (stable)

In the isotope generator, the cesium-137 radioactive atoms are bound on a special matrix. When a washing solution (eluent) is forced through the generator, the product (barium-137m) is washed off the matrix and collected in the washing solution (eluate). As the barium-137m decays to its ground state by emitting gamma radiation, it forms the stable barium-137 isotope.

Gloves and lab coats should be worn when working with all liquid radioisotopes.

As always, wash your hands thoroughly before leaving the lab, and then check for possible contamination. The final product of this decay is stable and can be safely washed away.

 ISBN 978 1 4886 1936 6

PRACTICAL ACTIVITY 8.3

You will use a G–M counter to monitor the decay of the barium-137m in order to calculate the decay constant and half-life of the isotope. The G–M counter will provide the rate-of-decay data (the number of atoms that decayed during one second).

1 Connect the G–M tube to the counter and set the stopwatch to zero. Alternatively, if using a data-collection system, follow the manufacturer's instructions to connect the G–M tube and prepare for data collection. Create a graph of counts per second versus time.

2 Collect the barium-137m solution from your teacher on a small aluminum plate. Place it in front of the G–M tube and start the counter and stopwatch or data-collection system immediately.

3 Collect data for approximately 20 min, manually recording the count initially every 30 s and then every 60 s. Record the data in Table 1 in the Data and analysis section.

DATA AND ANALYSIS

Use this table only if you are sampling manually. There's no need to transfer your results if using a data-collection system. Complete the data table by calculating ln (counts s^{-1}).

TABLE 1 Counts versus time for barium-137m decay

Time (s)	Counts	Counts/sec	ln (counts s^{-1})
0	0	0	
30			
60			
90			
120			
180			
240			
300			
360			
420			
480			
540			
600			
660			
720			
780			
840			
900			
960			
1020			
1080			
1140			
1200			

PRACTICAL ACTIVITY 8.3

1 Plot a graph of counts per second versus time in the space below or print and paste in the graph from your data-collection system.

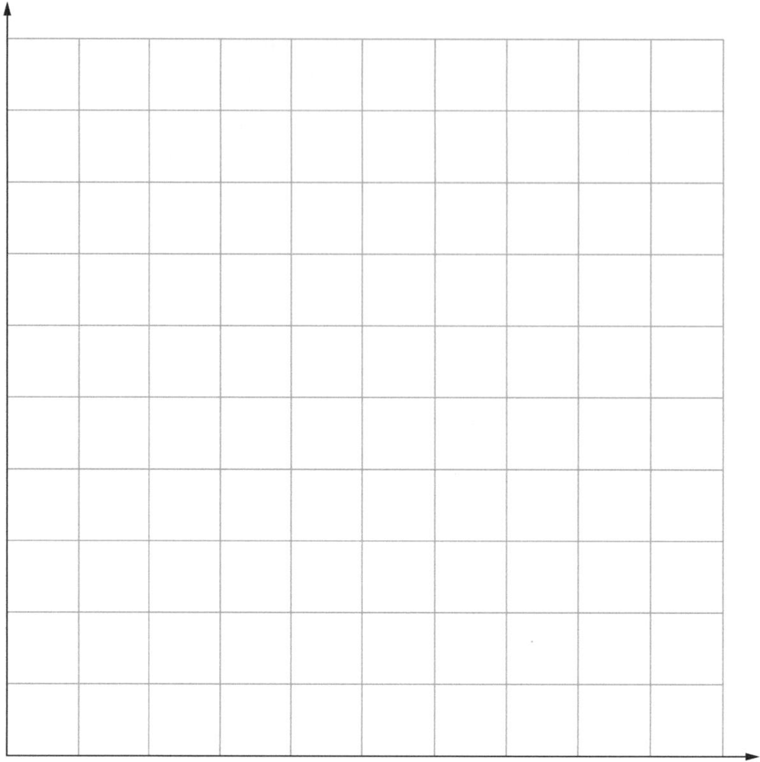

2 Plot a graph of ln (counts s^{-1}) versus time or print and paste in the same graph from your data-collection system.

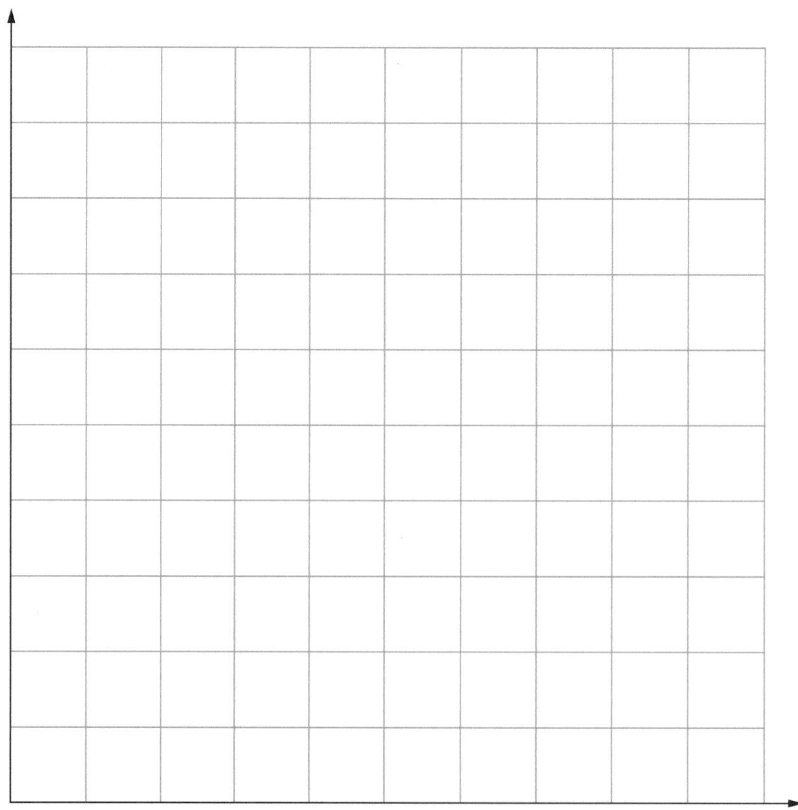

3 Why did you have to start the data recording immediately?

ISBN 978 1 4886 1936 6

PRACTICAL ACTIVITY 8.3

4 What happened to the count rate after 20 min?

5 Find the gradient and y-intercept of the graph. What quantity is the gradient of this graph?

6 Calculate the initial amount of barium-137m. Remember that the activity is given as $\frac{\Delta N}{\Delta t} = -\lambda N$, so the y-intercept of the natural log of the activity will be ln (λN_0).

7 Calculate the half-life of barium-137.

8 Find the theoretical value of the half-life of barium-137m and calculate the percentage difference between the theoretical value and the practical value you've found as a measure of the reliability of your result.

$$\% \text{ difference} = \frac{\text{theoretical value} - \text{experimental value}}{\text{theoretical value}} \times 100$$

Summarise the values you have found in Table 2.

TABLE 2 Determination of the half-life of barium-137m

Parameter	Value
λ (s^{-1})	
y-intercept	
N_0	
experimental $t_{1/2}$ (s)	
theoretical $t_{1/2}$ (s)	
difference (%)	

CONCLUSION

1 Comment on the value you found and its precision compared with the theoretical value.

2 What does it mean to say that the half-life of the barium-137m isotope is a particular number of seconds?

PRACTICAL ACTIVITY 8.3

3 How would you determine N_0 from the ln N_t versus t graph?

4 How could you determine N_0 from the rate of decay versus time graph? (Hint: Each point represents the number of radioactive atoms decayed during that minute.)

5 With reference to the percentage difference found between theoretical and practical values for the half-life of barium-137, how reliable is this practical method in determining the half-life of a radioactive isotope?

RATING MY LEARNING	My understanding improved	Not confident ⟷ Very confident ○ ○ ○ ○ ○	I answered questions without help	Not confident ⟷ Very confident ○ ○ ○ ○ ○	I corrected my errors without help	Not confident ⟷ Very confident ○ ○ ○ ○ ○

 ISBN 978 1 4886 1936 6

DEPTH STUDY 8.1

The OPAL research reactor, Lucas Heights—how can the energy of the atomic nucleus be harnessed?

Suggested duration: 3.75 hours

INTRODUCTION

Australian Government

ansto

Nuclear-based science benefiting all Australians

Beginning in the late 19th and early 20th centuries, experimental discoveries revolutionised the accepted understanding of the nature of matter on an atomic scale. The work of Thompson, Rutherford, Bohr, de Broglie and others developed a more complex model on which the experimental investigations of the late 20th and early 21st centuries built our current understanding of the atomic model.

This depth study requires you to conduct an investigation to collect valid and reliable data and information on the operation of the OPAL research reactor located at the Australian Nuclear Science and Technology Organisation, Lucas Heights, Sydney. You will develop an inquiry question that requires research to develop an informed hypothesis, plan a research or experimental investigation, analyse secondary data and information from appropriate sources, and use problem-solving techniques to determine the validity of the data or sources. You will evaluate secondary data to form conclusions by considering the quality of the data. You will select and process the data and information in order to communicate your findings in a fact sheet of approximately 500 words. The fact sheet should communicate scientific understanding using suitable language and terminology, scientific notation and nomenclature, and graphs and diagrams for a member of the general public to understand the development of the topic. The fact sheet should include evidence-based arguments supporting the role of OPAL, developed following peer evaluation of your arguments and conclusions.

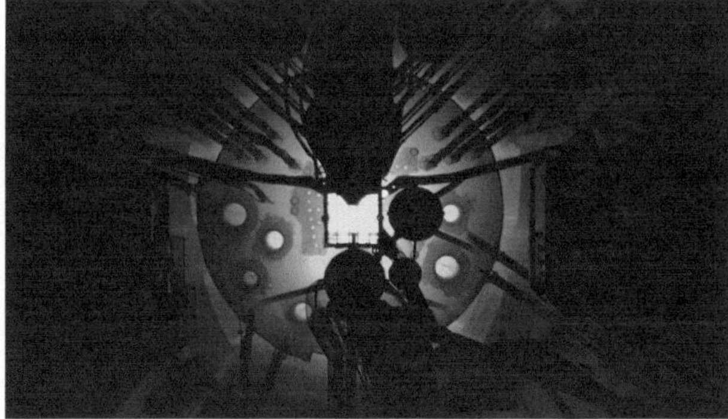

FIGURE 8.13 Inside the OPAL research reactor. The core containing the reactor fuel can be seen in the centre, with ports for irradiating targets surrounding the core.

PURPOSE

Research an aspect of the development, operation or function of the OPAL research reactor and report on your knowledge and understanding of the topic based on an analysis of secondary-sourced data. Your final analysis should be peer reviewed and adapted based on feedback to ensure clear communication of your conclusions.

TOPIC REQUIREMENTS AND CONSTRAINTS

Your research must be conducted individually.

The topic chosen must relate to the OPAL research reactor or the related technologies employed by or supported by this research reactor.

The topic must allow for the development and answer of one clear inquiry question related to the reactor and the supported technologies.

QUESTIONING AND PREDICTING

1 What is OPAL? What general capabilities does it have?

2 Phrase your topic as a question. For example, 'How do the control plates control the fission reaction in OPAL?' Some suitable starting points in the process of controlled fission in the OPAL research reactor include:

- the composition of the reactor fuel and the benefits of using low-enriched uranium (LEU)
- the structure of the core of the OPAL reactor and of an individual fuel assembly with a discussion of how structure relates to function
- the role of the control plates in controlling the fission reaction in OPAL
- the characteristics of hafnium that make it a suitable material for use in control plates
- the products of the fission reaction in the OPAL reactor
- neutron production in the fission reaction and the use of neutrons for different purposes in the OPAL reactor:
  - production of radioisotopes
  - diffraction research
  - silicon doping
  - neutron activation analysis
- the safety of the OPAL reactor, including structures and processes in place for protecting people and the environment from radiation produced by fission and fission products.
- compare the structure and function of the OPAL reactor with an average power reactor.

3 List three or four possible questions that would help answer your main question. For example, 'What do the control plates absorb?', 'What other structures assist in the control of the fission reaction?', 'How are the control plates controlled?'

4 Do some initial research around answering your initial questions. You may want to rephrase your questions to make them as clear and specific as possible. List your references and rephrased questions in the space provided.

 ISBN 978 1 4886 1936 6

5 Look back at your initial topic question. Construct a hypothesis that can be applied to answer your question. This becomes your working hypothesis and should summarise the answer to your main question. It may change after some further research.

CONDUCTING YOUR INVESTIGATION

6 ANSTO makes many resources available to the public, particularly for educational purposes. Start your research by visiting the ANSTO website and making appropriate contact with ANSTO. Identify the resources available for your topic through your own review—do not expect ANSTO to identify them for you. Confirm the steps you have taken to ensure the validity of the secondary-sourced data as it applies to your hypothesis.

7 Working in small groups, evaluate other students' topics. What are the strengths of their experimental investigations or research? How could you improve your own investigation or research question?

PROCESSING DATA AND INFORMATION

8 Carry out the remainder of your investigation, summarising key points as you proceed so that your topic has a clear focus and development. Use the tree diagram below to link your development back to your original hypothesis.

ANALYSING DATA AND INFORMATION

Evaluate your conclusions using the following questions.

9 Were you successful in answering your original hypothesis? Did you need to rephrase it?

10 Do your research and secondary data sources support your conclusions? How reliable is the data?

COMMUNICATING

Communicate your findings in the form of a fact sheet. It should include explanations of the topic development, assumptions you have made, relevant calculations, data, graphs and diagrams along with suitable explanation for the intended audience.

11 Review other students' research. Are they presenting evidence-based arguments in a form readily understood by the intended audience?

 ISBN 978 1 4886 1936 6

MODULE 8 • REVIEW QUESTIONS

Multiple choice

1 A Hertzsprung–Russell diagram plots stellar:

A redshift vs distance.
B size vs colour.
C colour vs brightness.
D brightness vs temperature.

2 The quark composition of a neutron is:

A uud.
B udd.
C du.
D uu.

3 If the three types of nuclear radiation are arranged in order of increasing ionising power, which of the following is correct?

A gamma, alpha, beta
B alpha, beta, gamma
C gamma, beta, alpha
D beta, gamma, alpha

4 According to quantum mechanics, all particles can behave like waves. The wavelength of a particle:

A decreases with bigger mass and bigger velocity.
B increases with bigger mass and bigger velocity.
C decreases with bigger mass and increases with bigger velocity.
D increases with bigger mass and decreases with bigger velocity.

5 Hadrons include:

A protons, muons and electrons.
B neutrons, protons and photons.
C protons, electrons and photons.
D protons, neutrons and muons.

6 The expansion of the universe was discovered by analysing:

A H–R diagrams.
B galactic redshift data.
C cosmic microwave background radiation.
D muon decay.

Short answer

7 a Given Planck's constant, $h = 6.63 \times 10^{-34}\,m^2\,kg\,s^{-1}$, calculate the wavelength of a 700 g grapefruit thrown at $18.0\,m\,s^{-1}$.

b Explain how your answer indicates that grapefruit do not seem to behave like waves.

8 Technetium-99m has a half-life of 6.0 hours and decays to technetium-99, emitting a gamma-ray photon of energy 140.5 keV. The equation for the decay is: ${}^{99m}_{43}\text{Tc} \rightarrow {}^{99}_{43}\text{Tc} + \lambda + 140.5\,\text{keV}$.

a Virtually all of the energy released during the decay process goes to the gamma photon. Keeping conservation of momentum in mind, explain why this happens.

b The gamma decay is followed by a beta decay to a stable isotope of ruthenium (Ru). Write the equation for this further decay.

c How long does it take for technetium-99m to reach 1% of its original activity?

9 A proton 'gun' accelerates protons into the first stage of a cyclotron by means of a potential difference of 4.50 keV. The protons enter the cyclotron following a circular arc with a radius of curvature of 6.0 cm. ($m_p = 1.67 \times 10^{-27}\,kg$, $q_p = 1.60 \times 10^{-19}\,C$)

a Find the speed at which the protons enter the cyclotron.

b Determine the strength of the magnetic field in the cyclotron.

10 The Rydberg formula, $\frac{1}{\lambda}=R\left[\frac{1}{n_f^2}-\frac{1}{n_i^2}\right]$, describes the wavelengths emitted from hydrogen atoms in a discharge tube.

a Explain what the quantities on the right-hand side of the formula represent.

b If the shortest wavelength emitted from hydrogen is 91.2 nm, find the ionisation energy for hydrogen.

c Use this information to find the value of the Rydberg constant, R.

d Which transition to the ground state gives rise to a wavelength of 93.8 nm?

11 One of the stages of the CNO cycle which takes place in stars is represented by the equation:

$$^{12}_{6}\mathrm{C} + ^{1}_{1}\mathrm{p} \rightarrow ^{13}_{7}\mathrm{N} + \gamma$$

Use the following table of data, where 1 amu = 1 u = 1.660539040 × 10^{-27} kg.

Particle	Mass (u)
$^{12}_{6}\mathrm{C}$	12.00000
$^{13}_{7}\mathrm{N}$	13.00573
$^{1}_{1}\mathrm{p}$	1.00728
$^{1}_{1}\mathrm{n}$	1.00866

a Find the mass defect in this reaction (in u and kg).

b Use the mass defect to determine the amount of energy released (in both J and MeV).

c Find the total mass of six neutrons and six protons (in u).

d What is the difference between this and the mass of a carbon-12 nucleus?

e Explain this difference and the energy represented by it.

12 Grypho-8, a Cepheid variable star, is observed to have a period of 4.8 days. It is in a galaxy 78 Mpc distant from the Earth.

a Why are Cepheid variables known as 'standard candles'?

b Alphonse-76c, another Cepheid variable with almost exactly the same period, is observed to be 225 times dimmer than Grypho-8. How far away is this star?

 ISBN 978 1 4886 1936 6

c The light from Alphonse-76c displays more redshift than Grypho-8. Is this observation consistent with the notion of the expansion of the universe?

d It is calculated that the galaxy containing Grypho-8 is receding from the Earth at a speed of 5.6×10^3 km s^{-1}. Use this data to find an estimate of the Hubble constant, H_0.

e Now find the recessional velocity of the galaxy containing Alphonse-76c.

ISBN 978 1 4886 1936 6

Formulae sheet

$$s = ut + \frac{1}{2}at^2$$

$$v = u + at$$

$$v = u^2 + 2as$$

$$\vec{F}_{\text{net}} = m\vec{a}$$

$$\vec{f}_{\text{friction}} = \vec{F}_{\text{N}}$$

$$\vec{p} = m\vec{v}$$

$$W = F_{\parallel}s = Fs\cos\theta$$

$$\Delta U = mg\Delta h$$

$$K = \frac{1}{2}mv^2$$

$$P = F_{\parallel}v = Fv\cos\theta$$

$$a_{\text{c}} = \frac{v^2}{r}$$

$$F_{\text{c}} = \frac{mv^2}{r}$$

$$\omega = \frac{\Delta\theta}{t}$$

$$\tau = r_{\perp}F = rF\sin\theta$$

$$v = f\lambda$$

$$d\sin\theta = m\lambda$$

$$I = I_{\text{max}}\cos^2\theta$$

$$I_1r_1^2 = I_2r_2^2$$

$$n_1\sin\theta_1 = n_2\sin\theta_2$$

$$f' = f\frac{(v_{\text{wave}} + v_{\text{observer}})}{(v_{\text{wave}} - v_{\text{source}})}$$

$$\lambda_{\text{max}} = \frac{b}{T}$$

$$F = \frac{GMm}{r^2}$$

$$g = \frac{GM}{r^2}$$

$$v = \frac{2\pi r}{T}$$

$$\frac{r^3}{T^2} = \frac{GM}{4\pi^2}$$

$$v_{\text{esc}} = \sqrt{\frac{2GM}{r}}$$

$$U = -\frac{GMm}{r}$$

$$E = -\frac{GMm}{2r}$$

$$t = \frac{t_0}{\sqrt{1 - \frac{v^2}{c^2}}}$$

$$\ell = \ell_0\sqrt{1 - \frac{v^2}{c^2}}$$

$$p_v = \frac{m_0v}{\sqrt{1 - \frac{v^2}{c^2}}}$$

$$E = mc^2$$

$$K_{\text{max}} = hf - \phi$$

$$E = hf = \frac{hc}{\lambda}$$

$$\frac{1}{\lambda} = R\left[\frac{1}{n_{\text{f}}^2} - \frac{1}{n_{\text{i}}^2}\right]$$

$$\lambda = \frac{h}{mv}$$

$$E = \frac{V}{d}$$

$$\vec{F} = q\vec{E}$$

$$F = qv_{\perp}B = qvB\sin\theta$$

$$F = \ell I_{\perp}B = \ell IB\sin\theta$$

$$\frac{F}{\ell} = \frac{\mu_0}{2\pi}\frac{I_1I_2}{r}$$

$$\Phi = B_{\parallel}A = BA\cos\theta$$

$$\varepsilon = -N\frac{\Delta\Phi}{\Delta t}$$

$$\frac{V_{\text{p}}}{V_{\text{s}}} = \frac{N_{\text{p}}}{N_{\text{s}}}$$

$$V_{\text{p}}I_{\text{p}} = V_{\text{s}}I_{\text{s}}$$

$$\tau = nIA_{\perp}B = nIAB\sin\theta$$

$$B = \frac{\mu_0 I}{2\pi r}$$

$$B = \frac{\mu_0 NI}{L}$$

$$V = IR$$

$$P = VI$$

$$E = Pt$$

$$N_{\text{t}} = N_0e^{-\lambda t}$$

$$\lambda = \frac{\ln 2}{t_{1/2}}$$

$$Q = mc\Delta T$$

$$\frac{Q}{t} = \frac{kA\Delta T}{d}$$

ISBN 978 1 4886 1936 6